AF764042

Environmental Technology Resources

HANDBOOK

Daniel W. Gottlieb

Environmental Technology Resources

HANDBOOK

LEWIS PUBLISHERS

A CRC Press Company

Boca Raton London New York Washington, D.C.

Library of Congress Cataloging-in-Publication Data

Gottlieb, Daniel W.
Environmental technology resources handbook/Daniel W. Gottlieb
p. cm.
Includes index.
ISBN 1-56670-566-5 (alk. paper)
1. Environmental protection--Information resources--Handbooks, manuals, etc. 2. Environmental engineering--Information resources--Handbooks, manuals, etc. 3. Pollution control industry--Information resources--Handbooks, manuals, etc. 4. Hazardous waste management industry--Information resources--Handbooks, manuals, etc. 5. Technology transfer--Information resources--Handbooks, manuals, etc. 6. Environmental protection--Directories. 7. Environmental engineering--Directories. 8. Pollution control industry--Directories. 9. Hazardous waste management industry--Directories. 10. Technology transfer--Directories. I. Title.

TD169.5 .G68 2002
363.7'0025—dc21 2002072485

This book contains information obtained from authentic and highly regarded sources. Reprinted material is quoted with permission, and sources are indicated. A wide variety of references are listed. Reasonable efforts have been made to publish reliable data and information, but the author and the publisher cannot assume responsibility for the validity of all materials or for the consequences of their use.

Neither this book nor any part may be reproduced or transmitted in any form or by any means, electronic or mechanical, including photocopying, microfilming, and recording, or by any information storage or retrieval system, without prior permission in writing from the publisher.

The consent of CRC Press LLC does not extend to copying for general distribution, for promotion, for creating new works, or for resale. Specific permission must be obtained in writing from CRC Press LLC for such copying.

Direct all inquiries to CRC Press LLC, 2000 N.W. Corporate Blvd., Boca Raton, Florida 33431.

Trademark Notice: Product or corporate names may be trademarks or registered trademarks, and are used only for identification and explanation, without intent to infringe.

Visit the CRC Press Web site at www.crcpress.com

© 2003 by CRC Press LLC

No claim to original U.S. Government works
International Standard Book Number 1-56670-566-5
Library of Congress Card Number 2002072485
Printed in the United States of America 1 2 3 4 5 6 7 8 9 0
Printed on acid-free paper

Cover art by Judy Smith

Dedication

To Diane

…Fire, plant and mineral say, "Not in us,"
And haughtily return us stare for stare.
For we invade them impiously for gain…
Therefore they shove us from them, yield to us
Only what to our griping toil is due;
But the sweet affluence of love and song,
The rich results of the divine consents
Of man and earth, of world beloved and lover,
The nectar and ambrosia, are withheld
And in the midst of spoils and slaves, we thieves
And pirates of the universe, shut out
Daily to a more thin and outward rind,
Turn pale and starve.

—Ralph Waldo Emerson, *Blight*

Preface

Environmental technology has been defined in many different ways. One of the broader definitions from a White House document characterizes it as any technology that:

- reduces human and ecological risks, enhances cost effectiveness,
- improves process efficiency and creates products and processes
- that are environmentally beneficial or benign.*

The Commerce Department gives an industry definition of environmental technologies to include air-, water- and soil-pollution control, solid- and toxic-waste management; site remediation; engineering, design and consulting services; environmental monitoring; recycling and industrial and clean-process technology.

For purposes of inclusion in this book, the definition used—to more precisely circumscribe the choices of what was to be included—is technology for environmental solutions (TFES).

TFES includes tools, materials, processes directed at identifying, quantifying, characterizing, minimizing, abating, remediating or preventing pollution or disturbances of nature's life support systems.

It is also important to recognize that the process of application of TFES or any other technology (oversimplified by the term technology transfer) depends on supportive or implementing mechanisms. These mechanisms may involve more general technologies (e.g., information technology, statistical controls and process engineering), economic principles such as market incentives, and mathematical tools such as modeling.

Regardless of the exact definition, the user or developer of environmental technology faces a myriad of sources for locating an appropriate technology or supporting mechanism for a particular environmental solution. This handbook serves as a guide for searching more efficiently the thousands of government agencies and programs, international bodies, local authorities and nonprofit and professional organizations that offer information in the public domain about environmental technology.

To save the reader time in browsing through an organization's Web site or publications, the handbook offers the reader a number of filters. For each organizational listing, for example, the reader will find key words, the "main resources" offered and the "technology functions" that are served (see Chapter 3 for a fuller description of these filters). In this way the reader can search, for example, only organizations that provide funding for development or application of TFES or only those that have site remediation technologies, or by a specific technology such as anaerobic decomposition.

In considering the particular purpose of an application for a TFES, the handbook groups them into four principal functional categories:

- Control—includes "end-of-pipe" treatment, disposal and neutralization of wastes and effluents.

* National Science and Technology Council, *Bridge to a Sustainable Future*, Washington, 1995

- Remediation—includes removal or isolation of toxic contamination at sites and other measures to make the sites nonthreatening to neighboring habitats or suitable for redevelopment.
- Assessment—includes activities to measure the nature and extent of environmental threats or damage and to monitor the effectiveness of technology on environmental quality.
- Prevention—includes all pollution prevention (P2) activities and efforts to make operations, products and processes sustainable, such as recycling, remanufacturing, materials accounting, resource conservation or minimization, life cycle analysis (LCA) or life cycle costing (LCC), design for the environment (DFE) and industrial ecology.

A broader concept that has gained widespread currency but is not well defined is "sustainability." The implication it carries of zero waste, closed-loop recycling and a non-polluting society, while a worthy ideal, is not inherent in any particular environmental technology. Where organizations use sustainability to describe their mission or technologies, however, the handbook adopts the term as such. It is enclosed in quotes here to suggest that sustainable operations, systems, or processes—at the community, national, or global level—require work in all four of the principal categories mentioned above. Removing toxics from abandoned industrial sites, for example, is classified as remediation, but remediation may occur together with measures to return them to productive uses that actually add environmental value over the long term, such as parks and trees that absorb urban heat and carbon dioxide emissions.

Finally, it is obvious that, in the rapidly expanding and shifting world of environmental priorities and technological developments, the author is selective of content in order to make the handbook manageable. Hopefully, it will serve as a useful starting point to bring information on what is available to potential users, developers and vendors so that more technologies for environmental solutions realize their potential.

Acknowledgments

The author is indebted to many persons who have given encouragement and assistance in the development of this handbook. David Scott-Smith offered insightful and invaluable guidance through the labyrinth of U.S. federal government resources. As well, he has been that indispensable fellow believer in the value of the project. Various reviewers, including Alan Miller of the Global Environment Facility and Charles Davis of the Department of Energy, helped sharpen the focus of portions of the book. In the same category, special mention is due my editor David Packer, who patiently gave encouragement and discerning suggestions along the way. To Pat Roberson and Sylvia Wood of CRC Press, I am grateful for guidance through the technical production trials of contemporary publishing. I have also benefited from the research, editing, and drafting of two extremely intelligent and diligent editorial assistants, Kristen Pomponio and Heather King, who also helped organize and manage the large volume of electronic files and paper.

Author

Daniel W. Gottlieb is an independent writer and editor who has specialized in environmental, health and safety topics. His credits range from high-level government policy reports and national magazine articles to technical manuals. He writes a fortnightly Washington column and a monthly "Regwatch" roundup on environmental, health and safety regulation and compliance for *Purchasing* magazine. He has also been a contributor to *Hazardous Materials Transportation* and *Safety Compliance Letter.* His bylined magazine articles have appeared in *Safety+Health, Washington Post, National Journal* and *High Technology.* A former reporter for the Washington bureau of Reuters news agency, the *Hartford Times* and *Washington Evening Star,* he has covered environmental affairs at the international, federal, state and local levels. He holds an M.A. from Johns Hopkins University's School of Advanced International Studies.

Table of Contents

1

Forces Driving Environmental Technology

The emergence of environmental technology[1] is the result of a number of forces that began to emerge in the latter part of the 20th century. These forces, listed roughly in the order of their appearance, include:

- Public awareness
- Environmental regulation
- Green consumerism and investment
- Environmental leadership
- Voluntary standards
- Technology verification and user consortiums

The combined impact of these forces has greatly expanded the role of environmental technology from that of mitigating specific pollutants to one that requires a combination of technologies, incentives, and public cooperation and participation to make possible changes in societal processes and lifestyles that will reduce pressures on life-sustaining natural resources. This shift in environmental goals—in both the public and private sectors—is encapsulated in the concept of "sustainability" of the planet for future generations[2]—a focus that calls for action not just by government but by business, nonprofit organizations and citizens, either in cooperation with government or on their own.

The remainder of this chapter describes the factors driving environmental technology so that organizations and individuals can make better judgments over the long term regarding the feasibility or value of investing in a particular application.

Public Awareness

Since the early 1960s, public awareness of environmental and health risks from pollution and its effects on the quality of life has grown to be fairly universal. This widespread public consciousness of environmental values can be attributed to a number of developments: a series of environmental disasters; scientific evidence of gradual environmental deterioration; and public education by governments, schools and environmental advocacy groups.

Environmental Disasters

Over the last three decades, several major pollution disasters and events that received wide media coverage include the following:

- Love Canal, New York, the locale of an abandoned chemical dumpsite, which was found to be the cause of illness and odors that forced the evacuation of more than 500 homes. It was the first non-natural disaster to be declared by a president and contributed momentum to the passage of "Superfund" legislation.[3]
- Middle East oil embargo, which created shortages of petroleum worldwide, bringing home the finite nature of fossil fuel and how energy wasteful industrial countries had become.
- Bhopal chemical plant (Union Carbide) release in India, which galvanized concern about storage and handling of hazardous chemicals and led to passage of the Toxic Substances Control Act (TOSCA).
- Three Mile Island nuclear reactor leak in Pennsylvania and meltdown of the Chernobyl nuclear reactor in Russia, which together contributed to the pressure for finding less-polluting energy sources, such as solar and wind power.
- Exxon Valdez oil spill in Alaska, which led to tighter regulations for tankers and underground tanks.

Scientific Evidence

Another factor raising public consciousness has been scientific and policy research that has exposed the less visible and long-term effects of pollution and the pressures placed on earth's natural resources by population growth. Some significant milestones in the growing body of evidence include:

- Rachael Carson's *Silent Spring*. Published in 1962, Carson's work on the effects of pesticides on songbirds is regarded as the harbinger of the modern environmental movement.[4] It is widely credited with the subsequent banning of DDT and the movement toward natural pest control methods.
- The Club of Rome Report. This and other similar reports have raised questions about the ability of the planet to support continuing population growth and accumulation of material wealth.[5]
- The ozone hole. Depletion of the ozone layer that protects against the harmful wavelengths of solar radiation[6] led to adoption in 1987 of the Montreal Protocol banning certain ozone-depleting chemicals (see discussion of global agreements under the Regulation section that follows).
- Global warming. Although scientists differ on the relative contribution of human activity, increases in global temperatures since 1861 — with much of it occurring during the 20th century — have been documented,[7] along with other likely indicators of climate change such as rising sea levels, thinning ice caps and gradual shrinking of mountain snow.

Public Education

During the 1980s and 1990s, growing business resistance to regulation[8] spurred the rise of environmental advocacy groups and the introduction of environmental awareness and course materials in the public schools. Advocacy groups also raised public awareness through lobbying in Congress, disseminating information and pressuring business to disclose information on actual or threatened environmental risks from their operations.[9]

As a result of all these developments, a loose global network of business, nonprofit and citizen groups, termed the environmental or "green" movement has emerged. Its members range from green political parties in Germany and the United States to thousands of temporary and permanent organizations at the local level.[10]

Early in the new millennium, the social justice element of environmentalism has also become more prominent. The World Bank observes, for example, that 2.8 billion people who live on less than $2 per day are "disproportionately affected by bad environmental conditions and natural catastrophes."[11]

This "institutionalization" of the environment in the public and private sectors ultimately creates pressure for solutions, which, in turn, creates pressure for regulation and opportunities for development and deployment of environmental technology.

Regulation

While the current emphasis in United States government policy appears to be on voluntary and collaborative government–private actions to improve environmental quality (see discussion under Voluntary Standards in this chapter), regulation still remains the prime driver for innovative technology. Much of the pressure for compliance and voluntary action derives from environmental laws and implementing regulations promulgated in the United States during the past three decades. From 1975 to 1995, for example, the number of environmental laws escalated from about 30 to 120 (Figure 1.1).

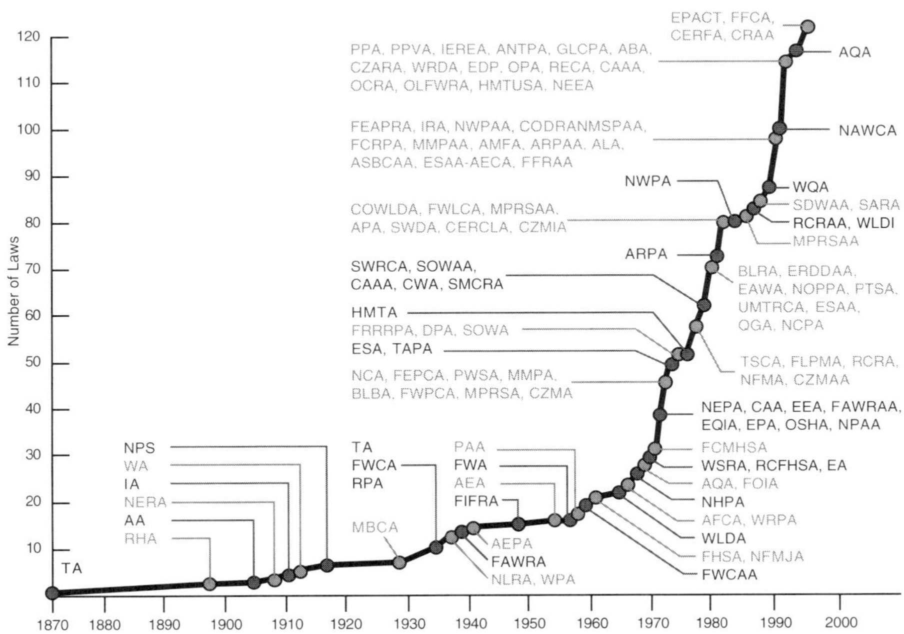

FIGURE 1.1
Growth of environmental regulation. Adapted from EPA chart.

Since the mid-1990s, however, the growth of new environmental legislation and regulation has slowed considerably as a result of a counter-regulatory campaign from business and some economists. In some cases, this counter force has led Congress and the federal government to pass laws and regulatory procedures that subject standards to the tests of cost–benefit, degree of risk, and scientific analysis, which have delayed and sometimes led to the abandonment or modification of proposed regulations. Still other factors are keeping the pressure from regulation alive.

Enforcement of Previously Enacted Laws and Regulations

Some statutes, particularly in the clean air and clean water areas, have compliance milestones extending throughout the current decade and beyond. Of the 1232 "superfund" sites listed by EPA as national priority for remediation, for example, slightly less than 60% (757) have been completed.[12] And many of the largest urban areas continue to violate ambient air quality standards, a problem that will be further aggravated if summer temperatures continue rising, which would increase emissions and pollutants due to greater energy production (for air conditioning) and more "smog" inversions (from vehicle emissions).

Broader Impacts of Newer Regulations

Some of the newer regulations, such as those affecting land use on or near wetlands and nonpoint-source standards for maximum daily loading on inland waterways, will soon bring regulatory impacts to a broader range of industries, farm producers and local governments than the specifically regulated industries in the past, such as chemicals, paper mills, metals finishing and utilities. These regulations likely will generate demand for different types of technology as well as for some that are already on the shelf.

Ripple Effects of Other National Regulations

Other industrialized or industrializing nations are moving in some areas toward higher standards of environmental protection than currently exist in the United States. For example, the European Union's Eco-label Award for personal computers goes beyond the energy-saving requirements of the U.S. Energy Star label. Also, personal computers sold in Europe must, in addition to meeting energy efficiency standards, have key components guaranteed to last a certain number of years, and the manufacturer must agree to take the entire PC back for recycling at the end of its useful life. And, beginning with the 2007 models, the European Commission has mandated that automobile manufacturers must cover the cost of recycling their vehicles and that at least 85% of the weight of each vehicle must be recyclable, as well as targeting waste electrical and electronic equipment (WEEE) for mandatory recycling and waste reduction.

In a world of rising trade volumes, such standards have ripple effects around the globe. Multinational and large U.S. manufacturers, at least, are tracking Europe and other regional organization directives very closely. Because multinational companies market their products globally, they generally find it more effective to bring their processes and products in line with the highest standards in the world rather than to balkanize their product lines to meet differing national standards.

Global Agreements

As national governments have discovered the transboundary and global presence of certain environmental problems, they have turned to regional and world organizations to set frameworks for collective action. Examples include the United Nations Framework Convention on Climate Change (Kyoto Treaty), the Convention on Biological Diversity and the Convention on Persistent Organic Pollutants. Nations and international lending organizations are developing policies and programs that, while not yet fully in effect (ratified), are intended to mitigate the problems they address. The Montreal Protocol on phasing out ozone-depleting substances is, to date, probably the most dramatic example of a global agreement that has driven development of environmental technology—for example, the incorporation into air-pressure spray cans of substitutes for chlorofluorocarbons.

Aside from specific environmental agreements, trade agreements are increasingly incorporating environmental provisions. One example is the North American Free Trade Agreement (NAFTA), which has set up a mechanism for dealing with transboundary pollution. Environmental groups are lobbying to direct the U.S. president to negotiate provisions in any future trade agreements that will require the parties involved to enforce their own environmental standards.

Green Consumerism and Investment

With broader awareness of pollution's effects, consumers have shown increased preference for environmentally friendly products. Many manufacturers and distributors, sensing competitive advantage, have been marketing such "green" products as consumer-waste recycled paper, recyclable packaging and electric power from renewable sources. Governments have also established environmentally preferable purchasing and certification programs. The U.S. government, for example, requires agencies to use paper with 30% post-consumer material and federal contractors to document that they have environmental policies and practices. Europeans have established eco-label certification standards for a range of products. Another manifestation of environmental consciousness is the emergence of the terms "green" and "socially responsible" in the 1970s. Additionally, some mutual funds screen the environmental behavior of companies to qualify them for inclusion in their portfolio. Such funds may become more than a "niche" market[13] and the value of the stock of socially screened companies has reached about $639 billion, according to one estimate.[14]

Environmental Leadership

Inspired by the thinking of seminal conservationists, many individuals have translated present environmental concerns into advocacy and action. The list is too numerous to cite in its entirety, but here are just a few:

- Amory Lovins has been named by *Wall Street Journal* editors as one of the persons most likely to change the face of world industry. In the 1970s, Lovens was credited with contributing significantly to shifting the energy debate from the supply to the demand side. He campaigned to convince utilities that it was good business to promote efficient energy use by their customers. Lovens received the 1997 Heinz Award for the Environment for his work in energy conservation and renewable technologies. He is co-founder, with his wife, of the Rocky Mountain Institute, which has developed potential pollution-prevention (P2) innovations such as a prototype ultra-lightweight, hybrid-electric vehicle. Two commercial auto manufacturers (Toyota and Honda) are already in the market with hybrids and several others are developing their own models.

- Ray C. Anderson, recipient of the 2001 George and Cynthia Mitchell International Prize for Sustainable Development and former chairman of the President's Council on Sustainable Development, has been called the "Green CEO" by *Fortune*. His company, Interface, the largest global carpet manufacturer, has launched more than 400 sustainability initiatives, such as designing carpets and fabrics that are 100% recyclable and a "climate-neutral" floor-covering product.
- When William D. Ruckelshaus, the first administrator of the Environmental Protection Agency (1970–73), returned to serve another term in the early 1980s, he initiated the process of seeking public consensus in rulemaking. It has since become more widely used in EPA rulemakings as well as by other regulatory agencies, although not always to the satisfaction of business or the environmentalists. In the 1990s, Ruckelshaus returned to warn Congress that cutting EPA's budget was not a solution to regulatory inflexibility and chaired a public–private-sector project that recommended greater use of economic incentives, emissions trading and information disclosure, an approach embraced by the current administration in Washington.[15]
- Taki Kiuchi, former chairman and CEO of Mitsubishi Electric America, became interested in how the rainforest, with poor soil, manages to flourish, and decided to promote it as a model for business operations. He currently leads a global business network, Future 500, which is dedicated to implementing technologies and policies that radically reduce their need for physical raw materials, from petroleum and pulp to metals and water.

In different ways, these and other environmental leaders have helped shape and redirect the environmental debate. Instead of pitting environmental advancement against jobs and profits, the debate now is about how to enhance all three. To put this into practice requires technology to eliminate toxic inputs, reduce resource inputs and recapture a greater percentage of the waste stream for reuse.

Voluntary Standards

In parallel with the forces of public awareness, regulation, green consumerism and environmental leadership, there has been a movement within the public and private sectors to adopt voluntary standards, some of which target goals beyond regulatory compliance.

Performance Standards

Regulatory agencies themselves have been shifting their focus from "command-and-control" and "end-of-pipe" regulation to more flexible performance standards,[16] thus increasing the opportunities for innovation in environmental technology. The current administration, for example, has proposed use of caps on emissions in combination with tradable pollution permits for power generators in achieving clean air standards for three air pollutants, as well as voluntary restrictions on greenhouse gas emissions.[17] These mechanisms are designed to encourage companies to find the most effective technologies for their individual operations rather than meeting "one-size-fits-all" solutions.

Individual and Collective Corporate Standards

Many large corporations have responded to the more flexible regulatory climate by increasing self-imposed standards.[18] In some cases, they have adopted voluntary standards promulgated by the International Standards Organization (ISO), such as the International Environmental Management Standard (ISO 14001). Initially, companies seeking ISO 14001 certification are required only to adopt environmental policies and management systems. The ultimate goal, however, is to use these systems to set environmental improvement targets and to meet them. To take these steps, however, management faces pressures from owners and shareholders to justify and prioritize environmental investments. The companies' challenge is to balance environmental investments among those required to protect their firms from compliance actions, those that achieve pollution prevention and those that add to bottom-line results. Figure 1.2 illustrates one prototype tool management can use to determine where in the product cycle environmental investments can add value to the business.

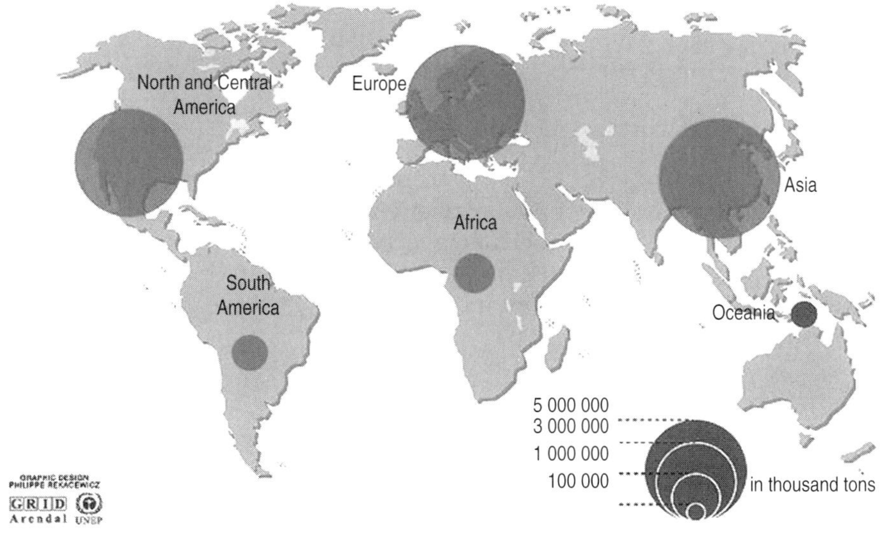

FIGURE 1.2
Growing carbon dioxide emissions from industry pose potential threats from global warming, such as climate change, draught, rising sea levels and loss of habitats for diverse species of flora and fauna. Credit: UNEP; design by Philippe Rekacewicz. Source: United Nations Framework Convention on Climate Change (UNFCCC).

Another development is called "greening the supply chain." Both government and business are setting environmental performance standards for the products and services they buy. As they do so, the smaller and medium-sized businesses that supply them will feel the pressure to "green" their operations. Ford and General Motors, for example, have required all their suppliers to adopt environmental management systems conforming to ISO 14001.

Collectively, businesses have also formed environmental alliances such as the World Business Council for Sustainable Development and the Global Environmental Management Initiative to collaborate in development of environmentally friendly practices.

Individually, some corporate leaders (see Anderson and Kiuchi, mentioned earlier, for example) have initiated programs to minimize total resource use and waste. The goal of achieving sustainable production is represented by a new corporate slogan, the "triple bottom line" (see Elkington) of environmental protection, profitability and social responsibility.[19] No company has claimed to achieve all three goals, but some have taken voluntary actions to benefit the environment that attempt to incorporate environmental value into business goals.[20]

Mitsubishi, for example, announced cancellation of a planned salt works project in Mexico because a scientific environmental-impact assessment showed that it would adversely affect a nearby lagoon's flora and fauna. Volvo is marketing an automobile with a radiator that cleans ozone from the air. Major chemical companies, in response to EPA requirements to report external releases of certain toxic chemicals, reduced those releases by 35% from 1988 to 1992 and the amount of wastes they ship off-site for treatment and disposal by 39%.[21]

Despite examples such as Volvo's ozone-eliminating radiator and other companies' billing products as environmentally friendly, the actual contribution to environmental improvement or the widespread melding of environmental and business values to achieve "sustainable" production is still uncertain.[22]

Technology Verification and User Consortiums

The emergence of technology verification and user consortiums is a further accelerating force for the development and deployment of environmental technology. The need to pool resources to develop, test and commercialize technology has brought more transparency to the market. Until recently, the dissemination of technology has largely been through the marketing efforts of vendors and informal technology transfer through environmental professional organizations.

Verification

The problem users face in selecting technology is that the environmental industry is highly fragmented, large and complex[23] and, in many instances, lacks approval from some standard-setting organization. It is divided into four primary segments: water supply and treatment, solid waste, air pollution, and a catchall for other environmental media. More recent niche-type services include conservation planning, natural resource damage assessment, and planning activities. Specialized technologies for use in remediation at Department of Defense or Energy Department facilities have also created another niche segment.

Some technologies are relatively mature and tested in the marketplace. For most, however, objective data on performance is lacking. This lack of independent and credible technology verification, according to EPA's science advisory board, has been "one critical barrier to the use of innovation technologies." Since 1995, however, EPA has addressed that barrier through its Environmental Technology Verification (ETV) program. California's Environmental Protection Agency and the Department of Defense offer similar ser-

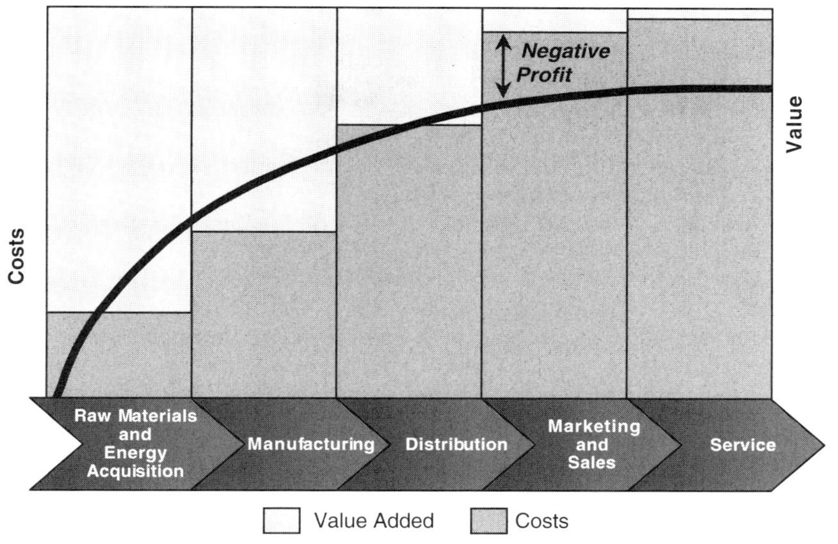

FIGURE 1.3
Investments in environmental technology, like other cost items, have differing values at particular stages of the manufacturing process. Illustrative chart shows opportunities are greatest at the raw materials stage and diminish as product moves into the market and is serviced. Credit: Robert D. Shelton, Arthur D. Little, Inc.

vices to environmental technology vendors (see entries in Chapters 5, 6 and 9 for details on these programs).

In response to the lack of verification and the confusing array of technologies available, some industry trade associations and business groups, as well as large and small urban communities, have joined consortiums — some in partnership with EPA and other government agencies — that pilot test technologies and share information on what works. A number of these are described in Chapters 4–9.

Conclusion

In this changing pattern of environmental technology development and dispersion, the search for environmental technology involves four challenges: identifying the environmental problem or challenge, knowing where to find relevant technology, obtaining the technical information needed for your particular application, and—after installation—having some means of knowing whether it is performing effectively. The next two chapters describe how to use this handbook to meet these challenges more effectively.

Endnotes

[1]See Preface for various definitions of the term and a definition adopted for this handbook.
[2]See Preface for a discussion of this concept.
[3]Comprehensive Environmental Response, Compensation and Liability Act (CERCLA).
[4]Carson, R. *Silent Spring,* Houghton Mifflin, Boston, (reprint edition), 1994.

[5]Meadows, D.H. *Limits to Growth: A Report for the Club of Rome's Project on the Predicament of Mankind.* Universe Books, Boston, 1972.

[6]This has been the result of international scientific endeavors over several decades, beginning with the work and inventions of a British scientist, James Lovelock, after World War II, and advancing with work on atmospheric chemistry by Sherwood Rowland and Mario Moline of the United States and Paul Crutzen of the Netherlands (who worked at the Max Planck Institute in Germany—all three sharing in the Nobel Prize in Chemistry in 1995—and aided by the work of Susan Solomon in Antarctica.

[7]See, for example, the Third Assessment Report of the Intergovernmental Panel on Climate Change, United Nations Environment Programme, October 2001. Available: http://www.ipcc.ch/

[8]The resistance came mainly from trade associations from affected industries and some economists, arguing that the costs of compliance would jeopardize economic growth.

[9]Despite relatively small budgets compared with industry associations and lobbies, their political influence and coverage by the media has been significant. One author describes them thus: "Despite setbacks in Congress, the environmental movement remains a powerful force in American politics … Environmental groups retain the power to frustrate industrial and commercial development and slow economic growth." Jonathan Adler, director of environmental studies at the Competitive Enterprise Institute.

[10]For a discussion of the impact of local environmental activism, see National Academy of Public Administration. *Resolving the Paradox of Environmental Protection*, Washington, DC, September 1997, pp.107–125.

[11]World Bank, Annual Report 2000,Washington, D.C. See section on Environmentally and Socially Sustainable Development.

[12]EPA data as of October 1, 2000. Updates available: http://www.epa.gov.

[13]See, for example, Green investment: will "real funds" break out of their niche markets? Kathy de Wit, paper for the 8th International Greening of Industry Network Conference, Chapel Hill, NC, November 14–17, 1999.

[14]Perryman, M. Ray, article in Houston business journal *American City Business Journals*, August 22, 1997 print edition.

[15]Enterprise for the Environment, Center for Strategic and International Studies, The Environmental Protection System in Transition: Toward a More Desirable Future, Washington, D.C. January 1998. Mr. Ruckelshaus now chairs Browning Ferris Industries and the board of the World Resources Institute.

[16]Performance standards set mandatory targets and timetables for pollution reduction but do not prescribe the technology to achieve them.

[17]See President George W. Bush's Clear Skies and Global Climate Change Initiatives, February 14, 2002.

[18]Multinational corporations, particularly, have adopted targets for pollution reduction and report on their progress toward achieving emission, energy, waste and material consumption reductions. See, for instance the annual reports of Coca Cola, General Motors, Weyerhauser, Kodak, Volvo and Mitsubishi, which post their environment performance reports on their Web sites.

[19]This phrase from Elkington's book is mirrored in corporate environmental reports. But Hoffman, A.J. in *From Heresy to Dogma: An Institutional History of Corporate Environmentalism,* New Lexington Press, San Francisco, 1997, suggests that the trends toward corporate environmentalism "are not driven by individual cost structures and attempts at greater efficiency but rather by the social influence of external actors." See particularly Chapter 7.

[20]See Environment: Value to the Top Line, Global Environmental Management Initiative, November 2001, Washington, D.C. Available: http://www.gemi.org.

[21]Chemical Manufacturers Association, A Pollution Prevention Report (1988-1992). Results are based on reports submitted by 1422 chemical plants for EPA's Toxics Release Inventory. Available at: http://es.epa.gov/program/regional/trade/cma-rprt.html.

[22]Hoffman, A.J. in *From Heresy to Dogma: An Institutional History of Corporate Environmentalism,* New Lexington Press, San Francisco, 1997. Hoffman argues (p. 197) that "there is no such thing as 'a green company' and that whether sustainable development becomes institutional corporate behavior depends on 'how society views the environmental issue and thus of how the organizational field defines the role of business in responding to it.' "

[23]This section is derived from the Department of Commerce's first comprehensive review of the environmental technology industry. See The McGraw-Hill Companies and U.S. Department of Commerce, *The U.S. Industry & Trade Outlook 2000,* Washington, D.C., Chapter 20.

2

How to Find TFES Resources

There is no shortage of information about environmentally friendly technologies, products and processes. A recent search on the Web for the entry "environmental technology" using multiple search engines yielded nearly 80 hits—and these were only the top ten selections from each Web search engine. Some offered multiple links to other sources. The plethora of sources and varied quality of the information on environmental technology make daunting the task of sorting out those that are effective and relevant to particular pollution problems.

The challenge is to be able to search *efficiently* and *effectively* for technologies that offer environmental solutions (TFES).

Think of the search process as a safari into a jungle to find a particular type plant. After cutting through brush for 2 days without encountering the plant, you send your native guide to climb up the tallest tree. After surveying in all directions, he shouts down: "Wrong jungle!!"[1]

Starting in the right jungle requires a similar overview. But it also requires knowing something about the different "species" of technologies." Thus, each search becomes like a new safari. This handbook will, however, take you directly to some targets. In addition, it will help you become a smarter scout in the jungle of environmental technologies.

Finding those that meet your needs, depends on four steps:

1. Focusing your search
2. Selecting the most likely source(s) (choosing the right jungle)
3. Identifying the assistance categories offered
4. Evaluating the information and its source (verification)

You may not always follow these steps in this order. You may want to perform a preliminary search — surf — over the jungle. This might open your mind to potential targets unforeseen when you first defined your search focus. But performing each step will increase your rate of success.

Focusing the Search

To avoid ending up in the wrong jungle, this step is crucial. As illustrated in Figure 2.1, you may decide to use such selection criteria as:

- Whether a specific technology is available commercially
- Which pollutant or affected environmental medium is involved

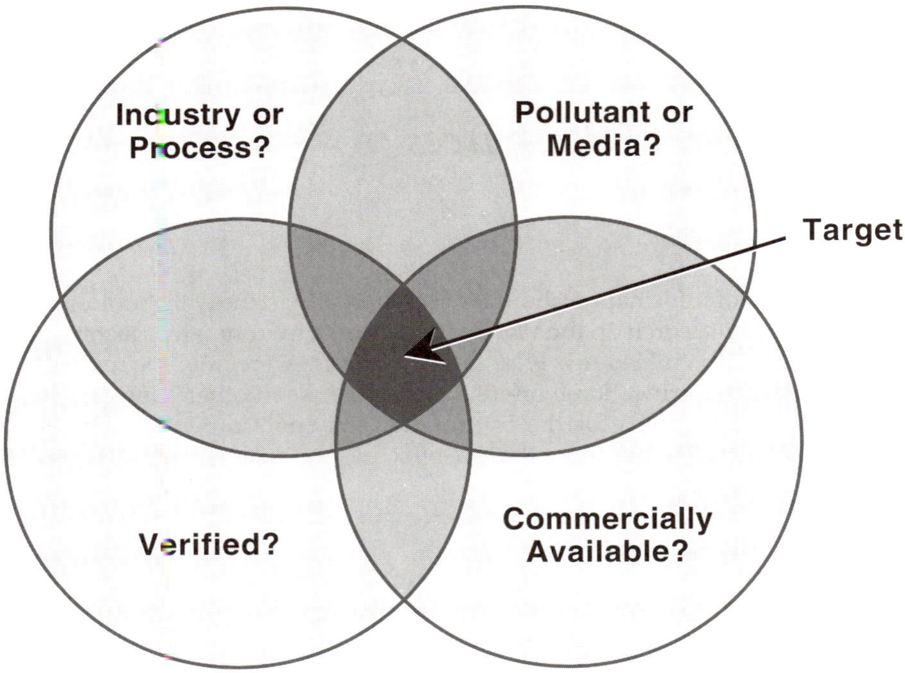

FIGURE 2.1
Selection criteria. Credit: Design by Stephen Hall.

- The industry or process (setting) in which the technology will be applied
- The results reported by users

Another way of narrowing the search is to decide what particular type of pollution remedy you are seeking: assessment, control, remediation, prevention, or sustainability.

No matter what the criterion, each can be described at differing levels of detail—from general to specific and from nontechnical to technical. You might be a printer wanting to know only about soy-based inks or press-cleansing solutions, for example. Or you might be a plant manager concerned with reducing a number of different waste streams, such as paper, lead batteries or water. The search focus will obviously be narrower in the case of the printer.

There is no general rule on just where to begin. Your initial target, for example, may be a technology that removes a specific airborne pollutant but that does not identify other possibilities that could be more cost effective. On the other hand, the site your search first brings up may have links that take you to them. A broader search might miss one particular technology but uncover one that removes the pollutant from water as well as the air. This could lead to a "two-for-one" solution.

Selecting the Source

Many organizations and programs offer TFES. Eight major types were screened for this handbook (for the selection criteria within each category, see Chapter 3):

1. Regulatory agencies
2. Government[2] technology development programs
3. Government technical assistance programs
4. Government grants and tax incentives
5. Government export promotion programs
6. Public–private partnerships
7. Nonprofit environmental organizations
8. International organizations

The handbook draws from all of these major resources for TFES. Some, such as trade associations, focus on a particular industry. Others focus on certain environmental media, ecosystems, or remedies relating to elements such as air, rainforests and renewable energy. Others mainly collect data and disseminate information. Then, there are those that provide grants, technical assistance and training for developing and adopting technologies.

Identifying the Assistance Categories

Government, business and nonprofit organizations usually have more than one category of assistance or information. Chapter 3 gives the taxonomy of assistance categories that are listed in the handbook.

The level of detail varies among the categories. Databases and manuals, for example, typically are very detailed. Technical briefs, on the other hand, tend to give more general descriptions of technology but usually have references to more detailed information. Market research reports generally are not fine tuned for a particular technology but for a class of products. In any case, you may want to follow up what you find on the Web or receive via e-mail or regular mail with e-mails or phone calls to obtain the detail you need. That is why the handbook offers contact information (Web url address, mail address, e-mail, phone, or fax, whenever it is available).

Evaluating the Information

Once you find a promising technology relevant to your needs, you probably want verification that it not only works but that it also is adaptable in the context in which you want to use it. The chief resources for evaluation (with examples of each in parentheses, and which are elaborated upon where they appear in the handbook) are described below:

Technology verification programs offer products that have been independently tested. (e.g., EPA's Environmental Technology Verification Program).

Case and pilot studies, if conducted or reviewed by outside experts, offer evidence that a technology has worked in a particular environment or application.

Success stories usually provide an account of results only by the developer of the technology or a single user or user group and the results may apply only in a particular instance. They can, however, provide leads to users from whom more detail can be obtained, such as the factors required for successful application of the technology elsewhere.

Training and technical courses may provide sufficient knowledge to evaluate whether a technology is suitable for the application intended. (United Nations Environmental Program).

Consortia formed by companies or groups within an industry or across multiple industries develop or refine technologies. They sometimes share information on results with outsiders (Global Environmental Management Initiative is a consortium of about 40 major corporations).

Public–private partnerships study, test and report on environmental technologies (EPA's Pesticide Environmental Stewardship, WasteWise, and Energy Star programs).

Federal or state technology extension services usually have experience with adapting technologies in varied settings (Manufacturing Extension Partnerships of the National Institute of Standards and Technology).

Research centers often seek partners in industry or government to participate in testing a concept or prototype in the partner's setting and evaluating the results thereof (Engineering Research Center for Environmentally Benign Semiconductor Manufacturing, National Science Foundation).

Chapter 3 provides an overview of how the handbook organizes the information to aid your search process.

Endnotes

[1]I am grateful to Stephen R. Covey for the anecdote from which this "jungle" description is adapted. See his *The 7 Habits of Highly Effective People.*

[2]Government covers federal, state and local and tribal units.

3

Mapping the Environmental Technology Jungle

Many organizations collect data about, study and report on the environment. It is more difficult to find those that actually develop or assist in transferring environmental technology. Government laboratories and research centers are a primary source, but many of them have broader mandates than environmental technology. To locate information on environmental technology *within* larger organizations, therefore, can be equally challenging.

The Selection Process

The handbook focuses on organizations that have a reputation for accuracy and are not highly specialized in one industry or technology. Necessarily, the choice of sources included is highly selective.

Six different criteria were used to select organizations with TFES resources:

1. Does the organization's *mission* include improving environmental quality?
2. Does it have *specific departments, branches* and *offices* with responsibilities for generating or collecting information on environmental technology?
3. Do its current *programs* address specific environmental media (e.g., air, water, solid waste)?
4. Do its *products* or *services* fall within major technology functions or applications (e.g., monitoring, energy efficiency, site remediation)?
5. Does it have a *technology transfer* (T2) activity designed to bring technology to the marketplace and users?
6. Does it have a *central contact point* or information management source?

When searching for government agencies with environmental programs, remember that the word *environment* does not necessarily appear in their titles, particularly in certain state and local governments. Some lodge environment programs within other departments or agencies, depending on the dominant industries within their jurisdiction. Thus, one may find environmental functions within agencies concerned with agriculture, conservation, forestry, parks, planning, public health, natural resources, mines, public works, or sanitation.

Refining Terms for Searches

Once the agency or organization (jungle) is located, the next challenge is drilling down to subdivision(s) with specific environmental technology responsibilities. If the Web site has a search engine, you might manage just by entering the term *environmental technology.* If you are looking for a particular type of resource, however, you will probably want to fine-tune your queries to obtain a more precisely relevant retrieval from the Web, or responses via any other means (fax, e-mail, letter, phone) by which you may be inquiring. The handbook offers two different matrices for refining the search: by the type of *organizational resources* offered and by the type of *technology functions.*

Organizational Resources

Most sources of TEES offer one or more types of organizational resources. They are sometimes labeled differently, but generally, they fall within these descriptions (some alternative descriptors are listed in parentheses):

R1—Case studies (project descriptions, results, success stories)

R2—Clearinghouses (databases, newsletters, e-mail notifications)

R3—Commercialization assistance

R4—Conferences (workshops, symposia, seminars)

R5—Consortia (joint development)

R6—Grants and cost sharing/(Cooperative Research and Development Agreements (CRADAs)

R7—Incentives (emission trading, tax credits, service pricing)

R8—Licensing (patents, intellectual property)

R9—Market intelligence (market assessments)

R10—Partnerships

R11—Software

R12—Technical assistance (outreach, technology transfer)

R13—Technology verification (field testing, demonstration)

R14—Training (education)

R15—Other (e.g., economic assessment, technology assessment, environmental measurements, benchmarking, environmentally friendly or substitute products)

The Technology Development Process

Sometimes it is difficult to locate these resources if they are not displayed on the organization chart or Web site buttons or pull-down menus. To know where to start drilling down, it is useful to understand the process of technology development and with which stage(s) of development an organization is associated. Briefly, those stages are:

- Basic research
- Applied research
- Development and refinement
- Commercialization
- Technology transfer

The last three categories are usually the most fruitful to explore, unless one is interested in new basic technologies being developed that might be incorporated into future environmental technologies.

The lines are not always sharply drawn between the various stages in the technology chain. Scientific laboratories or a government agency's external research program usually do not become involved in commercialization, for example. However, these basic and applied research organizations are probably the best source for finding promising potential technology spin-offs, for participating in joint research, or finding an opportunity for licensing a government-held patent. Conversely, the export assistance programs of the Commerce Department have little technical information but a considerable amount of market intelligence.

Perhaps the best place to start is by clicking on an organizational chart or site map link or button. In that way, you will have an overview and begin to locate where various responsibilities are pigeonholed. If you do not find specific information about grants or partnerships on the Web or in the literature of an organization, you will be able to identify the most likely person or unit that can provide it.

Technology Functions

If you are looking for technology by its type of use or function, the addition to your search of one or more of the following technology function terms may help (some terms used synonymously or that are contained within the principal technology function term are enclosed in parentheses):

T1—Analysis/Monitoring: (environmental impact statements, environmental measurement/metrics/surveillance)

T2—Clean Air/Climate Change (air pollution abatement, point-source and non-point-source pollution, greenhouse gases, carbon sequestration)

T3—Ecosystem Characterization/Management (eco-assessments, watershed management)

T4—Energy Conservation/Efficiency (energy management/analysis/rating)

T5—Environmental Management Systems (e.g., ISO 9000 and 14001)

T6—Process and product sustainability (design for the environment, industrial ecology, product stewardship, closed-loop systems)

T7—Recycling (resource recovery, reuse, remanufacturing)

T8—Renewable energy (solar, wind, biomass, geo- and ocean thermal energy)

T9—Renewable materials (biosolids from sewage sludge, biopolymers, industrial crops)

T10—Site remediation/hazardous wastes (abatement, enclosure, encapsulation, removal, treatment)

T11—Solid waste management (waste treatment, waste reduction, waste stream reduction)

T12—Wastewater/water quality (wastewater recycling, closed-loop systems, non-point-source runoff)

T13—Other

As new environmental problems come into focus, the technology terminology continues to expand. Concern over greenhouse gases (GHG), for example, is spawning research into carbon sequestration techniques. This may lead to soil-, crop- and forest-management technologies that result in greater terrestrial adsorption of GHGs.

As discussed in Chapter 1, environmental policy and technology are undergoing a paradigm shift. As the more obvious point sources of pollution are eliminated or reduced, policymakers, regulators, and concerned citizens and organizations need to broaden the range of technologies they seek. The handbook helps guide that process and provides information to enhance the rate of application of existing technologies.

Chapters 4–7 present resources within the U.S. federal government; Chapter 8, those in major international and regional organizations; Chapter 9, U.S. state and local programs and Chapter 10, major nongovernmental organizations (nonprofits). Each highlights the organization's main environmental technology subdivisions and programs, its environmental mission (purpose), the main organization resources and technology functions and points of contact for further information. Illustrative resources or technologies are also included where available.

4

Interagency Programs

PROGRAMS

- Office of Science and Technology Policy (OSTP)
- Council on Environmental Quality (CEQ)
- Federal Laboratory Consortium (FLC)
- Federal Remediation Technology Roundtable (FRTR)
- Small Business Innovation Research Program (SBIRP)
- Small Business Technology Transfer Program (STTRP)
- U.S. Global Change Research Program (GCRP)
- National Biological Information Infrastructure (NBII)

SUMMARY The federal government has no single operating agency that coordinates the development or dissemination of environmental technology. An effort was begun during the administration of President Bill Clinton to develop a *National Environmental Technology Strategy.* An *Interagency Environmental Technology Office* within the Council on Environmental Quality (CEQ) was established, but at the time of this publication had no continuing charter or operating funds.[1] The agency with the exclusive mission of protecting the environment, of course, is the Environmental Protection Agency (EPA). However, it is primarily concerned with regulating and setting standards. Its efforts to foster development of environmental technology are largely based on voluntary cooperation of other agencies, sometimes formalized by joint agreements. This handbook presents about 150 programs, spread across agencies, that address some aspect of environmental technology as part of their missions. It is not exhaustive and, not surprisingly, there is some overlapping of effort. The agency with the broadest mission in the federal government is OSTP, but it is mainly concerned with research and development policy. OSTP policymaking does, however, provide environment technology users and developers with clues about future developments in the field.

Of the agencies or programs in the handbook, two with limited involvement with other agencies are the Strategic Environmental Research and Development Program (SERDP) and the Environmental Security Technology Certification Program (ESTCP), both of which are described in the Department of Defense section of Chapter 5. The agencies listed above are the only interagency programs that involve more than three agencies.

Office of Science and Technology Policy (OSTP)

WEB SITE www.ostp.gov

KEY WORDS *biodiversity, climate change, ecosystems, environmental assessment, environmental permitting, monitoring, renewable energy*

PURPOSE to provide the president with timely policy advice and to coordinate the science and technology investment.

SUMMARY The role that *OSTP* plays is greatly influenced by how each president views that role. Directors usually have some influence over the allocation of research and development fund requests in the president's budget. They also coordinate policy among agencies regarding major scientific research and developments in which the government has an interest. Some environmental areas OSTP has been addressing include: global climate change research, climate observations and modeling, regional and national assessment of climate change impacts and ozone depletion; technology for renewable energy and energy efficiency; ecosystem research, biodiversity and invasive and endangered species; oceans and coastal research, integrated environmental monitoring and the development of methods and measures to determine the status and trends of the nation's ecosystems. OSTP conducts many of its coordinating functions through the president's National Science and Technology Council, which consists of cabinet-level representatives from the major research funding agencies in the government.

MAIN RESOURCES OFFERED R15 (environmental policy)

TECHNOLOGY FUNCTIONS Various

SECTOR RELEVANCE Various

CONTACT

Office of Science and Technology Policy
Executive Office of the President
Washington, DC 20502
Tel.: 202-395-7347
Email: ostpinfo@ostp.eop.gov

White House Council on Environmental Quality (CEQ)

WEB SITE www.whitehouse.gov/ceq

KEY WORDS air quality, climate change, environmental policy, energy conservation, environmental permitting

PURPOSE to coordinate federal environmental efforts and work with agencies and other White House offices in the development of environmental policies and initiatives.

SUMMARY Until 1997, *CEQ* published an annual report on the state of the nation's environment. Past reports are available from the FirstGov Web site and contain environmental data on all media as well as special focused reports (e.g., in 1996, on American rivers). Congress relieved CEQ of the requirement for an annual report and, since then, CEQ has focused on supporting various White House policy initiatives with environmental implications. The current priorities of the CEQ include coordinating innovative regulatory programs, such as EPA's Project XL, and promoting flexibility in regulatory agencies to increase the use of environmental management systems and market-based approaches. A major new initiative is streamlining the permitting of energy producing facilities. By Executive Order 13212, the president established an interagency task force, headed by the chairman of CEQ for the purpose of accelerating the completion of energy-related projects (such as pipelines, conservation and electricity transmission) while maintaining safety, public health and environmental protections. The task force is also charged with helping agencies to set up mechanisms for coordinating federal, state, tribal and local permitting procedures in geographic areas where increased permitting activity is expected. Other priorities include climate change and developing air pollution legislation for the 21st century.

MAIN RESOURCES OFFERED R1, R2, R7

TECHNOLOGY FUNCTIONS T1, T2, T5

SECTOR RELEVANCE Various

CONTACT

CEQ
722 Jackson Place, N.W.
Washington, D.C. 20503
Tel: 202-395-5750

For past reports of CEQ
www.firstgov.gov
Type in Council of Environmental Quality in the search engine.

Federal Laboratory Consortium for Technology Transfer (FLC)

WEB SITE http://www.federallabs.org

KEY WORDS technology transfer

PURPOSE to promote and facilitate technical cooperation among the federal laboratories and America's businesses, academia, state and local governments and federal agencies.

SUMMARY The *Federal Laboratory Consortium for Technology Transfer (FLC)* links R&D results from a nationwide network of more than 700 major federal laboratories, centers and their parent agencies to the marketplace. The FLC site lists about 60 environmental technologies at various stages, from those available for patenting to those already commercialized. Examples include a software package for evaluating solvent substitutes, an *in situ* heat remediation technology and a recycling process for used motor oil containers (see sample technology, below). Links describe how to share in R&D, acquire the technology in the laboratories and apply for grants. One mechanism for participating is through a cooperative research and development agreement (CRADA). FLC offers a list of the laboratories open to CRADAs and the R&D opportunities for outside partners. A monthly newsletter, *Newsline*, is available in PDF form. The other FLC publications can be ordered, in most cases for a fee, from the FLC Management Office.

MAIN RESOURCES OFFERED R1, R2, R5, R6, R8, R10, R12

TECHNOLOGY FUNCTIONS T1, T10

SECTOR RELEVANCE Various

CONTACT

FLC Management Support Office
950 North Kings Highway Suite 208
Cherry Hill, NJ 08034
Tel: 856-667-7727
Email: flcmso@utrsmail.com

Sample Technology: Supercritical CO_2 Treatment of Mixed Oil and Plastic Waste

WEB SITE http://www.federallabs.org (go to Technologies, Environmental, Success Stories)

KEY WORDS petroleum, recycle/reuse and solid waste

PERFORMER/DEVELOPER Department of Energy's Kansas City Plant (KCP)

ENVIRONMENTAL CHALLENGE At least 2 billion motor oil containers are disposed of annually in the United States. These "empty" containers hold a residue of nearly 15 million gallons of oil, more than what was spilled in the 1989 Exxon Valdez tanker accident in Alaska. The containers themselves represent enough usable plastic to fill more than 1200 rail cars. Recycling them required development of a cost-effective means of separating the oil from the plastic.

PROJECT RESULTS KCP engineers developed a separation process involving use of liquid and supercritical forms of CO_2, which acts as a solvent to wash oil-contaminated plastic chips. The result was clean plastic chips and usable oil, thus making recycling possible with zero waste generation. Honeywell FM&T (which manages and operates KCP) patented the technology in 1998 and recently licensed it to ITec International Technologies in Cincinnati, Ohio. ITec plans to incorporate it into cleaning equipment and to market it to oil-bottle manufacturers and -recyclers worldwide. The engineering team at KCP received the Missouri Chamber of Commerce's 2000 Governor's Pollution Prevention Award for its work.

CONTACT

Tanya Snyder
Kansas City Plant
Tel.: 816-997-5937
Email: tsnyder@kcp.com

Federal Remediation Technology Roundtable (FRTR)

WEB SITE www.frtr.gov

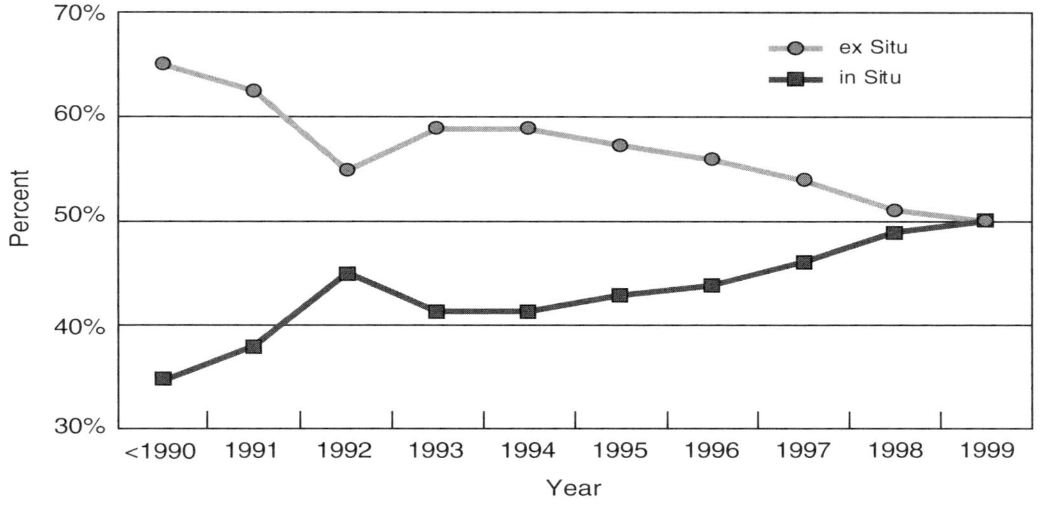

FIGURE 4.1
Use of *in situ* remediation is growing. Chart covers 230 case studies of soil and groundwater treatment that have been tracked from the year they were deployed. Source: Remediation Case Studies: Fact Sheet and Order Form at http://www.frtr.gov.

KEY WORDS remediation, site assessment

PURPOSE to provide information about, and links to, various state, federal, academic and commercial programs dealing with remediation technologies.

SUMMARY The *FRTR* is a portal to site-remediation technology programs in the state, federal, academic and commercial sectors. Most of the federal programs on the Web site are covered in the next chapter under the specific agency titles.[2] The Partnership and Consortia page links to some of the programs at each level, such as the federal *Innovative Treatment Remediation Demonstration Program (ITRD), Environmental Council of the States (ECOS)* and *Hazardous Substances Research Centers* in universities. The Technology InfoBase page lists both interagency and agency-specific programs, including a brief description, a contact person and a Web site link that lists 18 programs, although some, such as the Rapid Commercialization Initiative, no longer exist. A semiannual case study report has details on hundreds of site-remediation projects and can be ordered in hard copy. FRTR also lists upcoming regional meetings. The workgroups have various products to share and the site has a large alphabetic technology listing with links to detailed descriptions of each.

MAIN RESOURCES OFFERED R2, R4, R5, R10, R11, R12

TECHNOLOGY FUNCTIONS T10

SECTOR RELEVANCE federal government, state and local governments, various industries.

CONTACT For meetings and workgroups contact:

Chris Rossomando
Tel.: 301-589-5318
Email: chris.rossomando@emsus.com

Contacts for Programs can be found on the Contacts page.

Sample Resource: Summary of FRTR Case Studies – May 2001

WEB SITE http://www.frtr.gov/cost/index.html

The May 2001 report, available in PDF, documents 274 cases, including site background and setting, waste source, contaminants and media treated, technology design and operation, performance, cost, regulatory requirements, points of contact and lessons learned. See Table 4.1 for details on 56 sites.

TABLE 4.1

Summary of Remediation Case Studies

Site Name, State (Technology)	Principal Contaminants*: Chlorinated Solvents	BTEX and/or TPH	Pesticides/ Herbicides	MTBE	Metals	Radionuclides	Media (Quantity Treated)	Project Duration	Highlights
Other In Situ *Soil Treament*									
Alameda Point, CA (Electrokinetics)					•		Soil (38.4 m^3)	December 1997 to June 1998	Field demonstration of electrokinetic treatment of chromium in soil
Portsmouth Gaseous Diffusion Plant, X-231A Site, Piketon, OH (Hydraulic Fracturing)	•						Soil and Groundwater	1996 to 1998	Field demonstration of hydraulic fracturing with four types of remediation technologies
Incineration									
Drake Chemical Superfund Site, Operable Unit 3, Lock Haven, PA (Rotary Kiln Incineration)	•						Soil	January 1997 to April 1999	Use of on-site incineration to treat soil contaminated (273,509 tons) with VOCs and SVOCs, including herbicides
Thermal Desorption									
Metaltec/Aerosystems Superfund Site, Franklin Borough, NJ (Thermal Desorption)	•						Soil (4,215 yd^3)	December 1994 to January 1995	Use of thermal desorption to treat soil contaminated with chlorinated volatile organic compounds
New Bedford Harbor Superfund Site, New Bedford, MA (Thermal Desorption)							Sediment	November 1996	Demonstraton of thermal desorption/gas phase chemical reduction to treat PCB-contaminated sediments
Reich Farm, Pleasant Plains, NJ (Thermal Desorption)	•						Soil (14,836y^3)	November 1994 to March 1995	Thermal desorption of treatment of soils contaminated with VOCs an SVOCs
Rocky Flats Environmental Technology Site, Mound Site, Golden, CO (Thermal Desorption)	•						Soil (724.5 yd^3)	July to August 1997	Thermal desorption of soil contaminated with halogenated volatile organic compounds
Sarney Farm, Amenia, NY (Thermal Desorption)	•	•					Soil (10,514 tons)	August to December 1997	Thermal desorption treatment of soils contaminated with VOCs

TABLE 4.1 (CONTINUED)

Summary of Remediation Case Studies

Site Name, State (Technology)	Principal Contaminants*						Media (Quantity Treated)	Project Duration	Highlights
	Chlorinated Solvents	BTEX and/or TPH	Pesticides/ Herbicides	MTBE	Metals	Radionuclides			
Other Ex Situ *Soil Treatment*									
Brookhaven National Laboratory, NY (Physical Separation/Segmented Gate System)							Soil (625y^3)	May to June 2000	Use of a gate system to reduce volume of radioactive-contaminated soil requiring off-site disposal
Idaho National Environmental and Engineering Laboratory (INEEL), ID (Physical Separation/Segmented Gate System)						•	Soil (442 yd^3)	May to June 1999	Use of a gate system to reduce volume of radioactive-contaminated soil requiring off-site disposal
Massachusetts Military Reservation, Training Range and Impact Area, Cape Cod, MA (Solidification/Stabilization)					•		Soil (23,168 yd^3)	February to June 1998	Use of a proprietary stabilization technology to treat lead in both *in situ* and *ex situ* soils
New Bedford Harbor Superfund Site, New Bedford, MA (Solidification/Stabilization)							Sediment	November to December 1995	Bench-scale testing of solidification/stabilization to treat PCB-contaminated sediments
New Bedford Harbor Superfund Site, New Bedford, MA (Solvent Extraction)							Sediment	June 1996	Demonstration of solvent extraction/dechlorination to treat PCB-contaminated sediments
New Bedford Harbor Superfund Site, New Bedford, MA (Vitrification)							Sediment	July to August 1996	Demonstration of vitrification to treat PCB-contaminated sediments
Stauffer Chemical Company, Tampa, FL (Composting)			•				Soil (905 yd^3)	June 1997 to September 1998	Demonstration of composting technology for treatment of soil contaminated with chlorinated pesticides
Drinking Water Treatment									

Charnock Wellfield, Santa Monica, CA				•		Drinking Water	July 1998 to April 1999	Pilot-scale testing of Advanced Oxidation Processes (AOP) to treat MTBE and TBA in drinking water
Lacrosse, KS		•		•		Drinking Water	Ongoing, data from 1997 to September 1999	Use of air stripping to treat MTBE in drinking water
Rockaway, NJ	•			•		Drinking Water	Ongoing, data from 1980 to July 2000	Use of air stripping and GAC to treat MTBE in drinking water
Marine Corps Base, OU 1 and 2, Camp Lejeune, NC	•	•			•	Groundwater	Ongoing, data from January 1995 to March 1999	Use of pump and treat system to remediate groundwater contaminated with organics and metals
Marine Corps Base, Campbell Street Fuel Farm, Camp Lejeune, NC		•	•			Groundwater	Ongoing, data from 1996 to May 1999	Use of pump and treat system to remediate BTEX and SVOC groundwater contamination at three sites
Naval Air Station, Brunswick, Eastern Groundwater Plume, ME	•					Groundwater	Ongoing, data from May 1995 to May 1999	Use of pump and treat to treat groundwater contaminated with chlorinated VOCs
Ott/Story/Cordova Superfund Site, North Muskegon, MI	•		•		•	Groundwater (1.1 billion gallons)	Ongoing, data available from February 1996 to October 2000	Pump and treat of a multi-aquifer site contaminated with chlorinated and non-chlorinated VOCs and SVOCs
In Situ ***Groundwater Treatment***								
Brownfield Site, Chattanooga, TN (specific site name not identified) (Bioremediation)		•		•		Groundwater	Ongoing, data from January to December 1999	Use of *in situ* bioremediation to treat MTBE in groundwater
Butler Cleaners, Jacksonville, FL (Chemical Oxidation (KMnO4))	•					Groundwater	Not provided	Use of *in situ* oxidation with potassium permanganate to treat chlorinated solvents in groundwater at a dry cleaning site
Camp Lejeune Marine Corps Base, Bldg 25, Camp Lejeune, NC (*In situ* Flushing (SEAR and PITT))	•					Groundwater	Not provided	Use of surfactant injection to treat chlorinated solvents in groundwater at a dry cleaning site
Contemporary Cleaners, Orlando. FL (Bioremediation (HRC))	•					Groundwater	Not provided	Use of Hydrogen Release Compound (HRC) to treat chlorinated solvents in groundwater at a dry cleaning site

TABLE 4.1 (CONTINUED)

Summary of Remediation Case Studies

Site Name, State (Technology)	Principal Contaminants*: Chlorinated Solvents	BTEX and/or TPH	Pesticides/ Herbicides	MTBE	Metals	Radionuclides	Media (Quantity Treated)	Project Duration	Highlights
Cordray's Grocery, Ravenel, SC (Bioremediation (ORC®))		•		•			Groundwater	April 1998 to July 2000	Use of *in situ* bioremediation using ORC® to treat MTBE in groundwater
Eaddy Brothers, Hemingway, SC (Air Sparging/SVE)		•		•			Groundwater	Ongoing, data from July 1999 to August 2000	Use of air sparging/soil vapor extraction to treat MTBE in soil and groundwater
Eight Service Stations, MD (specific sites not identified) (Multi-Phase Extraction)		•		•			Groundwater	1990 to 1997	Use of Soil Vapor Extraction and Groundwater Extraction to treat soil and groundwater contaminated with MTBE and BTEX
Fernald Environmental Management Project, OH (*In situ* Flushing)					•		Groundwater	September 1998 to September 1999	Field demonstration of *in situ* flushing as an enhancement to pump and treat technology
Former Nu Look One Hour Cleaners, Coral Springs, FL (Air Sparging/Recirculation Well)	•						Groundwater	Not provided	Use of proprietary in well stripping technology to treat chlorinated solvents in groundwater at a dry cleaning site
Former Sages Dry Cleaners, Jacksonville, FL (*In situ* Flushing (Ethanol Co-Solvent))	•						Groundwater	Not provided	Use of surfactant injection to treat chlorinated solvents in groundwater at a dry cleaning site
Four Service Stations (specific site names not identified) (Air Sparging)		•		•			Groundwater	1993 to 1995	Air sparging used to treat MTBE in groundwater
Gas Station, Cheshire, CT (specific site name not identified) (Bioremediation)		•		•			Groundwater	October 1997 to March 1999	Use of *in situ* bioremediation to treat MTBE in groundwater
Hayden Island Cleaners, Portland, OR (Bioremediation (HRC))	•						Groundwater	Not provided	Use of Hydrogen Release Compound (HRC) to treat chlorinated solvents in groundwater at a dry cleaning site

Johannsen Cleaners, Lebanon, OR (Multi-Phase Extraction)	•				Groundwater	Not provided	Use of multi-phase extraction to treat chlorinated solvents in groundwater at a dry cleaning site
Lawrence Livermore National Laboratory, CA (Bioremediation)			•		Groundwater	Not provided	Research on microbial organisms to degrade MTBE in soil and groundwater
Louisiana Army Ammunition Plant, LA (Monitored Natural Attenuation)					Groundwater	Not provided	Field demonstration of MNA for remediation of explosives
Miamisburg, OH (Air Sparging/SVE)	•				Groundwater	Ongoing, data from December 1997 to May 2000	Full-scale treatment of chlorinated solvents using air sparging and soil vapor extraction
Monticello Mill Tailings Site, Monticello, UT (Permeable Reactive Barrier)				•	Groundwater	June 1999	Demonstration of permeable reactive barrier to treat heavy metals in groundwater
Multiple Sites (Monitored Natural Attenuation of Chlorinated Solvents)	•				Groundwater	July 1993 to August 1999	Field demonstrations of monitored natural attenuation for chlorinated solvents in groundwater at multiple Air Force sites
Multiple Sites (Monitored Natural Attenuation of Petroleum Hydrocarbons)		•			Groundwater	July 1993 to December 1998	Field demonstrations of monitored natural attenuation for fuel hydrocarbons in groundwater at multiple Air Force sites
Multiple Sites (Bioslurping)		•			Groundwater	Not provided	Field demonstrations of bioslurping of LNAPL at multiple Air Force sites
Naval Air Station, Pensacola, FL (*In situ* Oxidation)	•				Groundwater	November 1998 to May 1999	Use of Fenton's Reagent to remediate chlorinated solvents in groundwater
Naval Submarine Base, Kings Bay, GA (*In situ* Oxidation)	•				Groundwater	February 1999	Use of Fenton's Reagent to remediate chlorinated solvents in groundwater
Scotchman #94, Florence, SC (Air Sparging, Pump and Treat)		•	•		Groundwater	Ongoing, data from April 1998 to March 2000	Use of multiphase extraction and air sparging/soil vapor extraction to treat MTBE in soil and groundwater
Service Station, CA (specific site name not identified) (Bioremediation (ORC®))		•	•		Groundwater	Not provided	Use of *in situ* bioremediation using Oxygen Release Compound (ORC®) to treat MTBE in groundwater
Service Station, Lake Geneva, WI (specific site name not identified) (Bioremediation (ORC®))		•	•		Groundwater	Not provided	Use of *in situ* bioremediation using ORC® to treat MTBE in groundwater
South Beach Marine, Hilton Head, SC (Bioremediation)		•	•		Groundwater	Ongoing, data from February 1999 to September 2000	Use of *in situ* bioremediation to treat MTBE in groundwater

TABLE 4.1 (CONTINUED)

Summary of Remediation Case Studies

Site Name, State (Technology)	Principal Contaminants*: Chlorinated Solvents	BTEX and/or TPH	Pesticides/ Herbicides	MTBE	Metals	Radionuclides	Media (Quantity Treated)	Project Duration	Highlights
South Prudence Bay Island Park, T-Dock Site, Portsmouth, RI (Biosparging)		•					Groundwater	October 1997 to February 2000	Use of biosparging to treat BTEX-contaminated groundwater at a relatively remote site without collecting or discharging treated water
Sparks Solvents/Fuel Site, Sparks, NV (Multi-Phase Extraction)	•	•		•			Groundwater	Ongoing, data from 1995 to August 1997	Use of multiphase extraction using fluidized bed reactor and granular activated carbon to treat MTBE in groundwater
Specific site name not identified (Bioremediation)				•			Groundwater	Not provided	Bench-scale testing of the Butane Biostimulation Technologies™ process to treat MTBE in groundwater
U.S. Navy Construction Battalion Center, Port Hueneme, CA (Bioremediation)				•			Groundwater	April to August 1998	Laboratory and field testing of *in situ* bioremediation using MC-100 to treat MTBE in groundwater
Vandenberg Air Force Base, Lompoc, CA (Bioremediation)				•			Groundwater	Ongoing, data from 1999	Use of *in situ* bioremediation to treat MTBE in groundwater
Containment									
Dover Air Force Base, Groundwater Remediation Field Laboratory National Test Site, Dover DE (Vertical Engineered Barrier)							Groundwater	October 1996 to September 1998	Demonstration of vertical engineered barrier using thin diaphragm walls
Sandia National Laboratory, Albuquerque, NM (Cap)							Soil	July 1995 to July 2000	Field demonstration of alternative landfill covers

* Principal contaminants are one or more specific constituents within the groups shown that were identified during site investigations.

Small Business Innovation Research Program (SBIR)

Small Business Technology Transfer Program (STTR)

WEB SITE www.sba.gov/sbir/

KEY WORDS technology transfer

PURPOSE to strengthen and expand the competitiveness of U.S. small high-technology research and development businesses in the federal marketplace and to assist in achieving the commercialization of the results of federal research and development programs.

SUMMARY *The Small Business Administration's Office of Technology* carries out its mandate through two grant programs. *SBIR* funds the critical startup and development stages and encourages the commercialization of the technology, product, or service. Each year, ten federal departments and agencies are required to reserve a portion of their R&D funds for small-business research grants. They include the departments of Agriculture, Commerce, Defense, Education, Energy, Health and Human Services and Transportation as well as the Environmental Protection Agency, National Aeronautics and Space Administration and National Science Foundation. Awards are made in two phases. In Phase I, qualified small businesses are eligible for startup awards of up to $100,000 for about 6 months to support exploration of the technical merit or feasibility of an idea or technology. Successful Phase I awardees are eligible for Phase II grants of up to $750,000 for as long as 2 years, during which they perform the R&D to evaluate the technology's commercialization potential. Awardees are then expected to proceed to commercialization using private sector funds.

STTR is a highly competitive program that reserves a specific percentage of federal R&D funding for awards to small businesses working with nonprofit research institutions. STTR combines the strengths of both entities by introducing entrepreneurial skills to high-tech research efforts. The resulting technologies and products are transferred from the laboratory to the marketplace. Each year, five federal departments and agencies are required to reserve a portion of their R&D funds for STTR awards: Defense, Energy, Health and Human Services, the National Aeronautics and Space Administration and National Science Foundation.

Phase I Awards for up to $100,000, lasting about 1 year, are for exploration of the scientific, technical and commercial feasibility of an idea or technology. Phase II awards of up to $500,000, for as long as 2 years, expand Phase I results and are to consider the commercial potential. In Phase III, the small business must find funding in the private sector or from a non-STTR federal agency.

In addition to small businesses and nonprofits, states and territories categorized as "underserved" by federal R&D programs can receive grants of up to $100,000. Agencies determine the amount based on $2 of federal funding for each $1 of nonfederal matching funds the state invests in providing technical assistance and outreach for the STTR program. SBA expects to award between 10 and 24 state cooperative agreements. Eligible states and territories include: Alaska, Arkansas, Delaware, District of Columbia, Hawaii, Idaho, Indiana, Iowa, Kentucky, Louisiana, Maine, Mississippi, Missouri, Montana, Nebraska, Nevada, North Dakota, Oklahoma, Puerto Rico, Rhode Island, South Carolina, South Dakota, Vermont and Wyoming.

In addition,the Small Business Administration offers a large Internet-based database called *Tech-Net*. Offered free to small-business registrants, it provides links to companies seeking small-business partners, small-business contractors and subcontractors, leading edge technology research, research partners (small businesses, universities, federal labs and nonprofit organizations), manufacturing centers and investment opportunities. Tech-Net is a cooperative effort of the National Institute of Standards and Technology's Advanced Technology and the Manufacturing Extension Partnership programs (see Chapter 5).

MAIN RESOURCES OFFERED R1, R2, R5, R6, R9, R12

TECHNOLOGY FUNCTIONS Various

SECTOR RELEVANCE nonprofits, small business, states

CONTACT

U.S. Small Business Administration
Office of Technology—Mail Code 6470
409 3rd Street, SW
Washington, DC 20416
Tel.: 202-205-645[illegible]
Fax: 202-205-775[illegible]
E-mail: robert.connolly@sba.gov

U.S. Global Change Research Program (USGCRP)

FIGURE 5.2
Nearly a dozen U.S. agencies that focus research on protecting Earth's life-sustaining resources are coordinated through the USGCRP.

Web Sites http://www.usgcrp.gov/ and http://www.gcrio.org

KEY WORDS biodiversity, climate change, greenhouse gases, invasive species, lakes and rivers, land use, oceans, weather forecasting

PURPOSE to provide a sound scientific understanding of the human and natural forces that influence the Earth's climate system decision-making on global change issues.

SUMMARY *USGCRP* is an 11-agency effort coordinated through the National Science and Technology Council (See OSTP) Committee on Environment and Natural Resources (CERN). Research areas covered include climate change, stratospheric ozone depletion, ecosystems and land use. Most of the work to date has focused on monitoring, data collection, analysis of climate trends and observed and potential impacts of global warming, such as on rainfall, snow, ocean levels and sea ice. Most of the technology used has been applied to monitoring and analysis, except for the development of substitute propellant and refrigerant gases to replace chlorofluorocarbons that deplete stratospheric ozone. In February 2002, the president launched further research initiatives as part of his Clear Skies and Global Climate Change Initiatives. The technology initiatives include studies on how carbon dioxide can be stored more effectively by changing land use practices and forest management as well as by capturing CO_2 emissions and storing them underground. Centralized access to some of the GCRP research and collateral materials is available online and in published reports from the Global Change Research Information Office (GCRIO). The purpose is to disseminate scientific research information available in the United States that would be useful in preventing, mitigating, or adapting to the effects of global change.

MAIN RESOURCES OFFERED R1, R2, R4, R5

TECHNOLOGY FUNCTIONS T1, T2, T3, T4, T8

SECTOR RELEVANCE agriculture, energy, environmental services, forestry, solid waste

CONTACT

GCRIO User Services
PO Box 1000 61 Route 9W
Palisades, New York 10964 USA
Tel.: 845-365-8930
Fax: 845-365-8922
Email: help@gcrio.org

The National Biological Information Infrastructure (NBII)

WEB SITE www.nbii.gov

KEY WORDS biodiversity, invasive species, land use, parks, taxonomic information, vegetation mapping

FIGURE 4.3
A federal interagency effort is tracking and mounting counter measures to invasive species such as kudzu (Pueraria lobata), which has spread rapidly over large areas of the southeastern United States, killing trees and other native vegetation. An exotic species native to Japan, it was once planted widely in the U.S. for erosion control. (Photo courtesy of Kerry O. Britton, USDA Forest Service).

PROGRAMS

- Gap Analysis
- Land Use History of North America (LUHNA)
- The USGS-NPS Vegetation Mapping Program

PURPOSE to provide increased access to data and information on the nation's biological resources.

SUMMARY Hosted by the U.S. Geological Service, *NBII* links diverse biological databases, information products and analytical tools maintained by NBII partners and other contributors in government agencies, academic institutions, nongovernment organizations and private industry. Resource managers, scientists, educators and the general public use the NBII. It has numerous databases with search capabilities related to the management, use, or conservation of the nation's biological resources, all accessible through the central *NBII Metadata Clearinghouse.* Six regional and two thematic information nodes are being developed. The *Gap Analysis Program* identifies gaps in the protection of biodiversity state by state. Detailed graphics show land cover according to dominant plant groups, distribution of vertebrate animal species and level of protection being afforded. *Land Use History* *of North America* (LUHNA) examines land cover change and land use since 1900. Written by researchers for a lay audience and peer-reviewed by independent scientists, the historical profiles can be used for assessing current environmental conditions and interpreting current trends. Teaching resources are also included. The *Vegetation Mapping Program*, a cooperative effort of the U.S. Geological Survey and the National Park Service, classifies and maps vegetation communities in more than 270 national park units across the United States. Entries are peer-reviewed and information includes spatially enabled digital products available on the Web.

MAIN RESOURCES OFFERED R1, R2, R4, R14

TECHNOLOGY FUNCTIONS T1, T3

SECTOR RELEVANCE academia, agriculture, land developers, park management, planning agencies

WEB SITE http://www.nbii.gov/contact/

NBII National Program Office
USGS Biological Invormatics Office
302 Nation Center
Reston, VA 20192
Tel: 703-648-NBII (6244)
FAX: 703-648-4224

Endnotes

[1]See *Bridge to a Sustainable Future*, National Science and Technology Council (Washington, D.C., April 1995). An Interagency Environmental Technologies Office was established with links to environmental technology developers and vendors. It developed a Web-based "Federal Roadmap" of programs (www.getf.org/ietotech) with a limited search capability, but the future of this effort in the current administration has not yet been determined, according to officials at OSTP.

[2]The agencies include: the departments of Agriculture, Commerce, (National Oceanic and Atmospheric Administration), Defense, Energy, Health and Human Services, (National Institutes of Health), Interior, (U.S. Geological Survey) and State, and the Environmental Protection Agency, National Aeronautics and Space Administration, National Science Foundation and the Smithsonian Institution.

5

Federal Agencies

Department of Commerce (DOC) Web Site

www.doc.gov

KEY WORDS atmosphere, coastal zones, manufacturing technology, product development

PURPOSE to provide effective management and stewardship of the nation's resources and assets to ensure sustainable economic opportunities—(part of DOC's mission statement).

SUMMARY *DOC* environment-related programs include stewardship of the nation's coast and atmosphere (NOAA), promoting development of environmental technologies and standards (NIST), gathering statistics and analyzing trends in environmental industries and markets (ITA), promoting U.S. environmental industries' exports (ITA) and disseminating technical information on environmental topics (NTIS).

FIGURE 5.1
The Commerce Department is home to several programs that assist development of environmental technologies at home and promote their export abroad.

PROGRAMS
- International Trade Administration (ITA)
- Trade Development Administration (TDA)
- Environmental Technologies Industries Office (ETI)
- National Institute of Standards and Technology (NIST)
- National Oceanic and Atmospheric Administration (NOAA)
- National Technical Information Service (NTIS)

MAIN RESOURCES OFFERED R1, R2, R4, R5, R6, R8, R9, R10, R11, R12, R13, R14, R15

TECHNOLOGY FUNCTIONS T1, T2, T3, T4, T6, T7, T8, T9, T12

SECTOR RELEVANCE Various

CONTACTS See Agency entries

International Trade Administration (ITA)

WEB SITE www.ita.doc.gov

KEY WORDS environmental industry, exports, renewable energy

PURPOSE to open markets for U.S. products and provide assistance and information to exporters.

PROGRAMS

- Environmental Technologies Industries Office (ETI)
- Trade Development Agency (TDA)

SUMMARY Environmental industries are a special target for export promotion, market intelligence and assistance programs, as more and more foreign nations are prioritizing wastewater treatment, air quality monitoring and hazardous-waste projects. Services provided by ETI include Trade Information, *Export Programs Guide*, National Trade Data Bank, *U.S. Exporters Yellow Pages*™, trade leads and financing, industry sector data, market development, list of industry analysts, U.S. Export Assistance Centers (USEACs) and trade events and missions

MAIN RESOURCES OFFERED R2, R4, R6, R9, R10

TECHNOLOGY FUNCTIONS T1, T2, T3, T4, T5, T6, T7, T8, T9, T10, T11, T12, T13

SECTOR RELEVANCE environmental engineering, instrumentation, water treatment, wastewater treatment

CONTACTS

Trade Information Center
Tel: 1 800-USA TRADE
E-mail: tic@ita.doc.gov

For export assistance:
The Advocacy Center
International Trade Administration
U.S. Department of Commerce
Washington, DC 20230
Tel: 202-482-3896
Fax: 202-482-3508

Environmental Technologies Industries Office (ETI)

WEB SITE: www.environment.ita.doc.gov

KEY WORDS environmental industry, renewable energy

PURPOSE to facilitate and increase exports of environmental technologies, goods and services by providing support and guidance to U.S. exporters.

SUMMARY *ETI* is the principal resource and key contact point within DOC for American environmental technology companies. In the year 2000, it assisted U.S. environmental firms in winning $184.5 million in business abroad. The office analyzes environmental market trends abroad, identifies export opportunities, disseminates trade leads, counsels environmental companies on export strategy, conducts trade promotion activities, develops trade policy positions in the environmental sector, participates in trade negotiations and consultations and advocates on behalf of U.S. exporters in particular transactions. ETI has links to environmental industry facts, trade statistics, trade associations, export financing and other resources. See Global Technology Network (www.usgtn.org) and *National Environmental Exporters Resource Guide* for state export assistance agencies. The *Environmental Export News,* a quarterly newsletter, is available online or by subscription.

MAIN RESOURCES OFFERED R2, R4, R6, R9, R10, R12

TECHNOLOGY FUNCTIONS T1, T7, T8, T9, T10, T11

SECTOR RELEVANCE Environmental equipment, environmental services

CONTACT See Web site for various contacts

Sample Resource: Aid-to-Trade: Opportunities for U.S. Environmental Technologies Exports

WEB SITE web.ita.doc.gov/ete/eteinfo.nsf/FinancingChaptersATT?Openview

SUMMARY This report lists a number of programs offered by USAID, ITA and EPA to help U.S. businesses enter new markets. These programs offer services such as information support, trade promotion events, identification of trade leads and selection of local partners.

In addition, the report describes existing and prospective opportunities for U.S. companies in five countries: Egypt, India, Mexico, the Philippines and Ukraine. In all of these countries, U.S.-funded environmental and energy programs serve as a major market driver and create competitive advantages for U.S. companies through technology transfer and direct equipment procurement.

CONTACT

For print copy, call:
Office of Environmental Technologies Industries
ITA
U.S. Department of Commerce
Washington, DC 20230
Tel: 202-482-5225

National Institute of Standards and Technology (NIST)

WEB SITE www.nist.gov/

FIGURE 5.2
NIST Headquarters in Gaithersburg, Maryland.

KEY WORDS air quality, aquaculture, biotechnology, chemicals, composites, fire fighting, forests, groundwater, heating and cooling, lakes and rivers, manufacturing, materials, measurement, soils, standards, urban planning

PURPOSE to strengthen the U.S. economy and improve the quality of life by working with industry to develop and apply technology, measurements and standards and to provide U.S. industry with the world's best technical infrastructure and return the best possible value to the economy and society.

PROGRAMS

- Advanced Technology Program (ATP)
- Manufacturing Extension Partnership (MEP)
- NIST Laboratories

SUMMARY Established in 1901 as the Bureau of Standards, the nation's laboratory for development of scientific and commercial measures and standards has expanded its mission over the years. With an annual budget of nearly $600 million, *NIST* has become in the last decade one of the federal government's leading technology transfer agents. NIST has more than 400 nonprofit MEP centers in the 50 states and territories, its own specialized research laboratories and various databases and outreach services. In addition, it runs an accreditation program for testing laboratories and offers licensing of NIST-held patents. NIST partners with industry and state business assistance centers to generate and disseminate innovative technologies. NIST carries out its mission through developing standard measures, weights and reference materials; partnering with industry, universities and states; conducting research and making research grants; licensing NIST-held patents; providing technical assistance, technology development and commercialization assistance; and distributing software and other information products.

MAIN RESOURCES OFFERED R1, R2, R3, R4, R5, R6, R8, R10, R11, R13

TECHNOLOGY FUNCTIONS T1, T2, T4, T5 T6, T7, T8. T9, T10, T11, T13 (biotechnology)

SECTOR RELEVANCE academia, agriculture, manufacturing

CONTACT

NIST
100 Bureau Drive, Stop 3460
Gaithersburg, MD 20899-3460

Public Inquiries Unit:
Tel: 301-975-NIST (6478)
TTY: 301-975-8295

Advanced Technology Program (ATP)

WEB SITE www.atp.nist.gov/

KEY WORDS air quality, building chemicals, electronics, materials, minerals, natural resources, refrigeration, urban planning

PURPOSE to bridge the gap between the research lab and the market place, stimulating prosperity through innovation.

FIGURE 5.3
Advanced Technology Program (ATP). Composite bridge beams are designed to last longer and reduce life-cycle costs.(Strongwell Corporation, with permission.)

PROGRAMS

- Economic Assessment Office (EAO)
- Chemistry and Life Sciences (CLS) Office
- Electronics and Photonics Technology Office (EPT)
- Information Technology and Applications Office (ITAO)

SUMMARY Through cost-sharing partnerships with the private sector, ATP makes early stage investments to accelerate the development of innovative technologies that promise significant commercial payoffs and widespread benefits for the nation. The EAO evaluates results in the ATP program. CLS works in two main areas: (1) chemical, fuel and other bulk processing and (2) health, agriculture and industrial biotechnology. ITAO works in information technology for such environmental areas as air quality systems and natural resources modeling.

MAIN RESOURCES OFFERED R1, R2, R3, R4, R5, R6, R8, R10, R11, R13, R14

TECHNOLOGY FUNCTIONS T1, T2, T4, T6, T7, T8, T9

SECTOR RELEVANCE information networks, healthcare, manufacturing, composite structures, materials processing, microelectronics, DNA diagnostics, vapor compression refrigeration technology, motor vehicle, photonics. (Note: Projects are not confined to these industries. These are taken from ATP's existing inventory of projects.)

CONTACT

National Institute of Standards and Technology
Advanced Technology Program
100 Bureau Drive, Stop 4701
Administration Building 101, Room. A407
Gaithersburg, MD 20899-4701
Tel: 800-ATP-FUND [(800) 287-3863]
Fax: 301-926-9524 or 301-590-3053
E-mail: atp@nist.gov

Sample Technology: High Volume Wastewater Treatment for Aquaculture

WEB SITE www.nist.gov/public_affairs/techbeat/tb2000_09.htm

KEY WORDS aerobic, aquaculture, carbon dioxide (CO_2), fish farming, fish hatcheries, nitrification, nitrogen, sewage, wetlands

FIGURE 5.4
Fish farm increased production, reduced water pollution. (Kent SeaTech Corp.)

PERFORMER/DEVELOPER Kent SeaTech, San Diego, CA. Tel: 858-452-5765

PROJECT OBJECTIVE to develop a lower cost wastewater treatment system for fish farms.

FINANCING ATP grant, $2 million; company, $2million.

ENVIRONMENTAL CHALLENGE to reduce pollutant discharge from aquaculture operations. Fish farming is one of the growing segments of agriculture, with approximately 5,000 land- and marine-based facilities located in every state and territory of the United States. In addition to its largely state-administered NPDES (National Permit Discharge Elimination System), EPA has begun collecting data to establish final national effluent limitations guidelines and standards for discharges from fish hatcheries and farms.[1] The environmental sustainability of fish farms is affected by the undesirable byproducts discharged from the ponds that can increase nutrient loadings in nearby bodies of water.[2]

PROJECT RESULTS Patents pending on three-stage wastewater treatment and recirculation system. The Kent SeaTech facility is now able to reuse 80% of the water in the facility, convert nitrates to nitrogen, control the water's pH level and remove more than 75% of the volatile suspended solids. By reducing CO_2 and other byproducts from the water, production increased 40% (to 4 million lbs. annually) of striped bass (rockfish) at a cost of only a fifth of that by conventional sewage treatment. The system also reduces hydraulic retention times to one-half day to one day, compared with typical wetland treatment requiring 7 to 21 days (*NIST Update*, Sept. 2000 and Kent SeaTech).

TECHNOLOGIES EMPLOYED introduction of another fish species (tilapia), which thrives on by-product solids; a nitrification reactor termed SMART™ (suspended media ammonia removal technology), an aerobic bacterial process; and a wetland impoundment that provides a final clean-up of metabolites by bacteria associated with plant life (NIST, Kent SeaTech).

DEVELOPER COMMENT "The ATP grant very definitely sped development of the program. The ATP is very faithful to the goals of the program and very consistent across all types of companies, whether a Kodak or a small company. They look at the merits and what you're bringing to the research. It's kind of unique. We are engaged in a second ATP project,"—Michael Massingill, executive vice president.

CONTACT

Kent SeaTech
San Diego, CA.
Tel: 858-452-5765

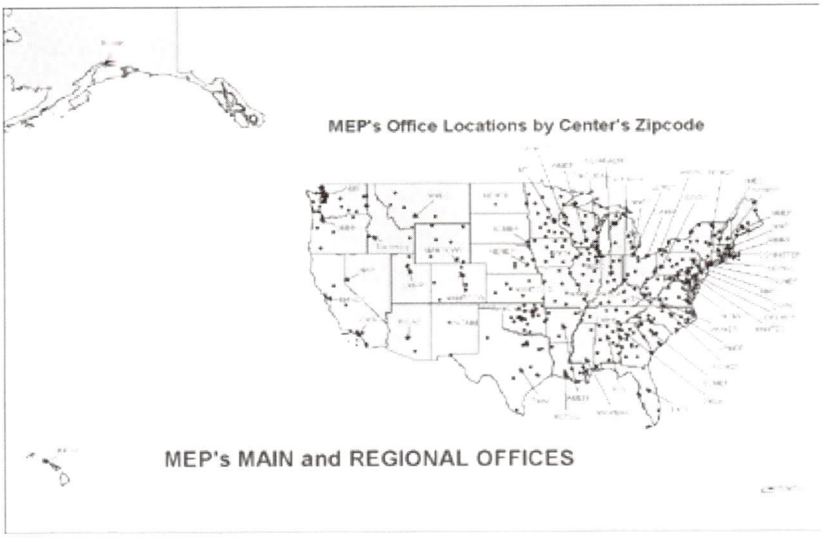

FIGURE 5.5
NIST manufacturing extension centers blanket continental United States, Alaska, Hawaii and Puerto Rico, offering assistance in productivity, process, and environment improvements to small and medium-sized companies.

Manufacturing Extension Partnership (MEP)

WEB SITE www.mep.nist.gov/

KEY WORDS energy audits, manufacturing technology, materials substitution, process improvement, product development

PURPOSE to provide small and medium-sized manufacturers (typically, those with fewer than 500 employees) with the help they need to succeed.

SUMMARY *MEP*, a nationwide network of not-for-profit centers in over 400 locations nationwide, serves all 50 states, the District of Columbia and Puerto Rico. Linked together, they enable even the smallest firms to tap into the expertise of knowledgeable manufacturing and business specialists all over the U.S. MEP customers typically seek assistance in locating appropriate resources or technologies, gaining expert impartial advice to help evaluate alternative solutions, determining the cause of product defects and implementing new technologies or processes to establish them as market leaders. The centers help form partnerships with other businesses and offer technical assistance, training and seminars. Areas of expertise include: process improvement, quality management systems, business management systems, human resource, development, market development, materials, engineering plant layout, product development energy audits, environmental studies, financial planning, CAD/CAM/CAE and electronic commerce/EDI.

MAIN RESOURCES OFFERED R1, R3, R8, R10, R12, R14
Technology Functions*: T4, T5, T6, T7, T10 (*Note: not limited to these, depending on MEP client needs.)

SECTOR RELEVANCE manufacturing

CONTACT

Tel: 800-MEP-4MFG (800-637-4634). Your call will be automatically routed to the MEP center that serves your region.

Sample Technology: Lead-Free Solder Consortium

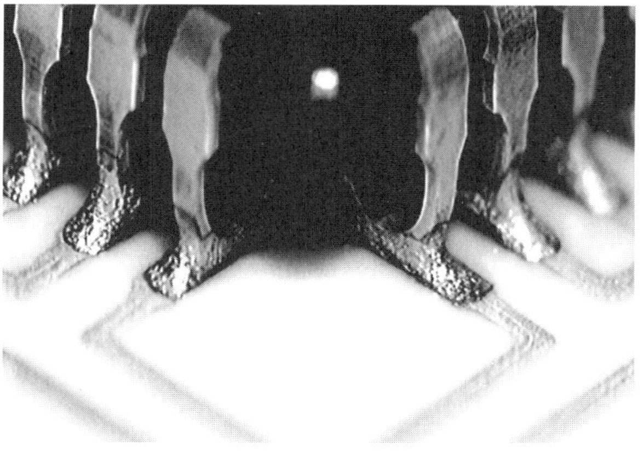

FIGURE 5.5A
Plastic leaded chip carriers on ceramic substrates assembled with lead-free solders and surface finishes. Photo courtesy of NCMS High Temperature Lead-Free Solder Project and NIST.

WEB SITES www.nemi.org/PbFree-PUBLIC/index.html
E-mail: www.nist.gov

KEY WORDS manufacturing technology

ENVIRONMENTAL CHALLENGE U.S. electronics companies are facing the need to find an alternative to lead-tin solder in manufacturing in order to meet a ban on lead in electronic equipment by Japan and the European Union, which is scheduled to take effect in the next few years. EPA is currently conducting a life-cycle analysis of products using lead-solder.

DESCRIPTION Solder is a critical part of assembly in circuit boards. NIST has been working with industry and university groups to develop a standard alloy that will serve as the alternative to lead-tin and ensure lead-free products. The National Electronics Manufacturing Initiative (NEMI), a consortium of electronics equipment manufacturers, government agencies and universities, launched the NEMI Lead-Free Assembly Project in 1999 to help North American companies develop the capability to produce lead-free products, with an eye toward total lead elimination by 2004. The consortium is now conducting manufacturing and reliability trials for lead-free circuit board assemblies. Meanwhile, NIST is developing a national lead-free electronics database on the properties of the lead-free alloy and on other types of solder materials. Preliminary tests indicate that, for some applications, the new solder offers greater reliability. At present, the data on the alternate solder is proprietary among the NEMI consortium members, but some technical information is available. See, for example, High Temperature Lead-Free Solders: Tools for the Microelectronics Industry, a NIST technical report at www.metallurgy.nist.gov/techactv2001/TechnicalHighlights.html#th_2.

SECTOR RELEVANCE electronics

CONTACT

Carol Handwerker
Chief of the Metallurgy Division
NIST
Tel: 301-975-6158.
E-mail: Carolh@nist.gov

NIST Laboratories

WEB SITE www.nist.gov/public_affairs/labs2.html

KEY WORDS ceramics, chemicals, information technology, manufacturing technology, measurements, metals, polymers, standards

PROGRAMS

- Building and Fire Research Laboratory (BFRL)
- Chemical Science and Technology Laboratory (CSTL)
- Electronics and Electrical Engineering Laboratory (EEEL)
- Information Technology Laboratory (ITL)
- Manufacturing Engineering Laboratory (MEL)
- Material Science and Engineering Laboratory (MSEL)
- Physics Laboratory (PL)
- Technology Services (TS)

FIGURE 5.6
The NIST Biomonitoring Specimen Bank© stores thousands of tissue samples for monitoring long-term pollution and health trends. (Robert Rathe, 1996, see Chemical Science and Technology Laboratory.)

PURPOSE to conduct research in a wide variety of physical and engineering sciences, responding to industry needs for measurement methods, tools, data and technology.

SUMMARY *NIST Laboratories* conduct research in a broad spectrum of physical and engineering sciences. NIST researchers collaborate with colleagues in industry, academic institutions and other government agencies. In addition to standards and measures in particular fields or industries, the labs help to develop innovations and bring them to market through Technology Services. The primary fields of research for each lab are described below.

BFRL works to improve quality and productivity in U.S. construction, including environmental performance (see Case Study). The lab also works to reduce human and economic loss due to fires, earthquakes, wind and other hazards.

CSTL conducts research in measurement science and develops the chemical, biochemical and chemical engineering measurements, data, models and reference standards required to enhance U.S. industrial competitiveness and to improve public health, safety and environmental quality.

EEEL provides the fundamental basis for all electrical measurements in the United States and metrology support to other federal and local government agencies.

ITL focuses on technologies to improve the usability, reliability and security of computers and computer networks for work and home.

MEL works to achieve greater efficiency and productivity with improved measurements and standards, both dimensional and mechanical.

MSEL provides expertise in ceramics, polymers, metallurgy, neutron characterization and materials reliability. It anticipates and responds to industry needs in areas such as microelectronics, automotive and healthcare, as well as providing standard reference materials and measurement methods. The lab houses the nation's only fully equipped cold neutron research facility.

PL provides measurement services and research for electronic, optical and radiation technology.

TS support U.S. industry to improve access to markets; disseminate information on weights and measures, calibrations and standard reference materials and data; train foreign standards officials; conduct laboratory accreditation; facilitate partnerships between NIST researchers and U.S. industry; and provide access to the NIST Research Library.

MAIN RESOURCES OFFERED R2, R3, R4, R5, R6, R8, R9, R10, R11, R13, R14

TECHNOLOGY FUNCTIONS including, but not limited to, T1, T6, T8

SECTOR RELEVANCE automotive, chemicals, computers, construction, electronics, health care, information technology

CONTACTS

See individual links on Web site for NIST lab contacts
Or contact:
NIST
100 Bureau Drive, Stop 3460
Gaithersburg, MD 20899-3460
Tel: 301-975-NIST (6478)
TTY: 301-975-8295
E-mail: inquiries@nist.gov

Inquiries about NIST Boulder Laboratories:
NIST-Boulder, MS 346.16
325 Broadway
Boulder, CO 80305-3328
Tel: 303-497-3198
TTY: 303-497-3665
E-mail: smithcn@boulder.nist.gov

Building and Fire Research Laboratory

WEB SITE www.bfrl.nist.gov/

KEY WORDS building materials, energy conservation, energy efficiency, fire fighting, manufacturing technology

PURPOSE to meet the measurement and standards needs of the building and fire safety communities.

PROGRAMS

- Enhanced Building Performance
- Office of Applied Economics

SUMMARY *The Building and Fire Research Laboratory* studies building materials; computer-integrated construction practices; fire science and fire safety engineering; and structural, mechanical and environmental engineering. It works with relevant industries to develop measurements, test methods, performance criteria and technical data that support innovations by industry. Its *Enhanced Building Performance* program develops measurement methods, fundamental data, simulation models and life-cycle environmental and economic analysis tools to support the widespread use of sustainability in design, construction, operation and demolition of buildings and their systems/subsystems. The *Office of Applied Economics* provides information to assist decision makers in choosing among new technologies relating to manufacturing, industrial processes, the environment, energy conservation, construction, facility maintenance, law enforcement and safety.

MAIN RESOURCES OFFERED R1, R2, R5, R11, R13, R14, R15

TECHNOLOGY FUNCTIONS T1, T4, T6

SECTOR RELEVANCE construction, fire-fighting equipment, heating and cooling

CONTACT

Building and Fire Research Laboratory
NIST
100 Bureau Drive, Stop 8600
Gaithersburg, MD 20899-8600
Tel: 301-975-5900
TTY: 301-975-8295

***Sample Technology: Building for Environmental and Economic Sustainability** (BEES™)*

WEB SITE www.bfrl.nist.gov/oae/software/bees.html

KEY WORDS building materials, energy efficiency, global warming, indoor air quality, life-cycle analysis, solid waste

ENVIRONMENTAL CHALLENGE to reduce indoor air pollution from building materials. Whether by architects, engineers or purchasers, decisions made on what materials and products to use can have major long-term impacts on the environment and life-cycle costs of buildings. The residential and commercial sector, for example, represents about 35.4%

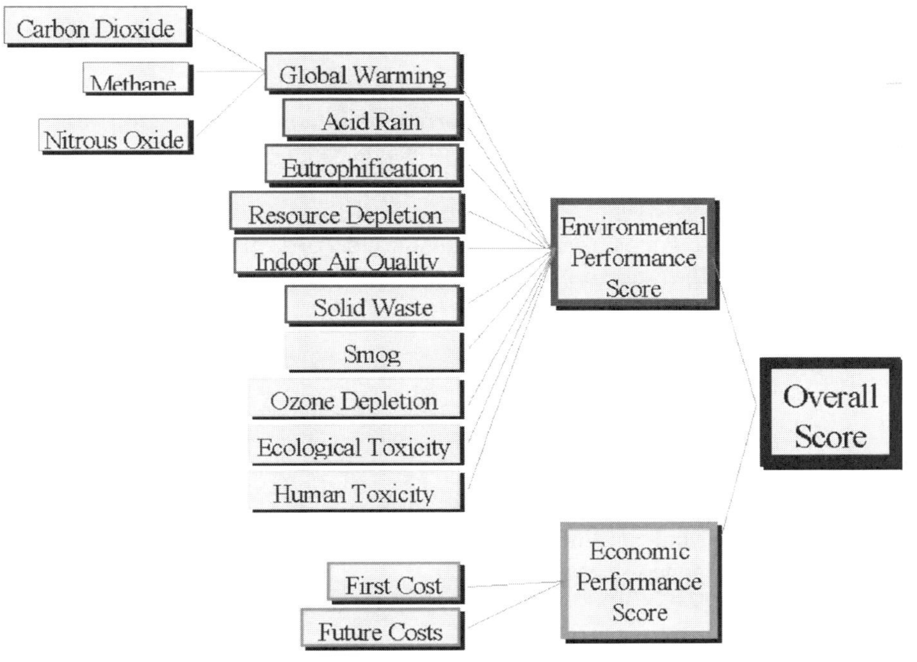

FIGURE 5.7
Building and Fire Research Laboratory (BFRL) BEES software measures tradeoffs between economic and environmental performance (NIST/Barbara Lippiatt, with permission).

of total U.S. energy consumption and it uses more than double the electricity that industry does.[3] Building materials and furnishings also affect indoor air quality. Indoors, where people spend as much s 90% of their time, pollution levels may be 2–5 times — and occasionally 100 times — higher than outdoors.[4]

PROJECT OBJECTIVE To provide an automated approach for assessing the life-cycle environmental and economic performance of building products.

PROJECT RESULTS BEES™ is freeware, a Web-based tool to measure and balance the environmental and economic performance of 65 products (more products are being added). It allows measurement of multiple environmental and economic impacts (global warming, resource depletion, first costs, future costs, etc.) over the entire life of the product. Developed by NIST in cooperation with EPA and scores of manufacturers and industry associations, more than 5500 copies of BEES 2.0 (3.0 is due for release in 2002) were requested by individuals from more than 65 countries in the first year it was available. NIST has a registry of users from industry, state and local government and nonprofit organizations.

TECHNOLOGY DESCRIPTION The software package allows users to compare generic product types (using an "industry average" for a dozen environmental and cost factors such as global warming, indoor air quality, solid waste, smog and future cost.[5]) The floor-covering product category, for example, includes ceramic and vinyl tile, carpets, marble and terrazzo. NIST is soliciting industry partners that wish to have their specific products evaluated. NIST lawyers are working on the issue of what participating manufacturers can advertise about their products, but NIST does not at present verify them, and manufacturers may not claim they are endorsed by NIST. To download BEES, go to download link at www.bfrl.nist.gov/oae/software/bees.html.

SECTOR RELEVANCE Architecture and engineering, building products, construction

USER COMMENT "BEES is the only tool in the United States that measures and balances life-cycle environmental and economic performance in a way we and our customers can understand ... BEES highlights these tradeoffs [between environmental and economic performance], and, in so doing, levels the playing field for all of us.[I]t is key that performance metrics be developed and delivered by an institution like NIST that is known for its credible, unbiased, science-based results."—David Ford, president and CEO, Certified Forest Products Council, letter to NIST, Oct. 17, 2000.

CONTACTS

Barbara C. Lippiatt
Office of Applied Economics
Building and Fire Research Laboratory
National Institute of Standards and Technology
100 Bureau Drive, Mail Stop 8603
Gaithersburg, MD 20899-8603
Tel: 301-975-6133
Fax: 301-975-5337

National Oceanic and Atmospheric Administration (NOAA)

FIGURE 5.8
Protecting coastal zone ecologies is one focus of NOAA's mission.

WEB SITE www.noaa.gov

KEY WORDS air quality, biodiversity, climate change, coastal zone, coasts, ecosystem, fisheries, lakes and rivers, protected species, stratosphere.

PURPOSE to predict environmental changes, protect life and property, provide decision makers with reliable scientific information and foster global environmental stewardship

PROGRAMS

- Coastal Zone Management Program (CZMP)
- Environmental Technology Laboratory/Air Resources Laboratory

SUMMARY NOAA's strategic goals include sustaining healthy coasts and monitoring air pollution. The CZMP assists state and territorial coastal management capabilities through financial assistance ($58 million in FY 2000), mediation, technical services and information, and participation in priority state, regional and local forums. NOAA maintains extensive databases on climate and other natural resources, such as the National Environmental Satellite, Data and Information Service (NEDIS), heating and cooling degree-day data in the monthly state, regional and national *Historical Climatological Series* (HCS) and *National*

Environmental Data Index, which provides distributed access to environmental information locator systems of several agencies. Its many laboratories, such as the ETL and ARL, provide the data and measurement techniques underpinning government environmental management programs.

MAIN RESOURCES OFFERED R2, R6, R8, R10, R12

TECHNOLOGY FUNCTIONS T1, T2, T3, T5

SECTOR RELEVANCE fisheries, offshore drilling, state and local government

CONTACT

Office of Public Affairs
14th Street & Constitution Avenue, NW
Room 6013
Washington, DC 20230
Tel: 202-482-6090
Fax: 202-482-3154
E-mail: answers@noaa.gov

Most NOAA photos and slides are in the public domain and cannot be copyrighted. Although, at present, no fee is charged for using the photos, credit *must* be given to the National Oceanic and Atmospheric Administration/Department of Commerce unless otherwise instructed to give credit to the photographer or other source.

Carla Wallace, Photo Administrator
NOAA Central Library
Tel: 301-713-2600 x118

Environmental Technology Laboratory (ETL)

WEB SITE www.etl.noaa.gov

Air Resources Laboratory (ARL)

WEB SITE www.arl.noaa.gov

KEY WORDS air quality monitoring, remote sensing

PURPOSE to meet environmental measurement challenges through theoretical and modeling studies; the design, development and testing of remote sensing instruments and techniques; conceptualization and execution of field programs and the acquisition and analysis of high resolution environmental datasets. ARL's mission is to improve understanding of processes that influence air quality, such as complex terrain, local meteorological conditions and long-range transport.

FIGURE 5.9
NOAA lab in the Houston area uses Doppler lidar to measure winds for better understanding of pollution transport. (NOAA Environmental Technology Laboratory, with permission.)

SUMMARY *ETL* researchers collaborate with colleagues around the world to create advanced remote sensors to meet environmental challenges. Remote sensing, the observation of the distant atmosphere using light, radio and sound waves, plays a critical role in weather forecasting, severe weather detection and warning, and air quality assessment. Ground, airborne and satellite systems collect high-resolution datasets necessary for improved modeling and forecasting of the environment. Doppler lidars and radars developed by ETL have been used in mesoscale pollution studies. ETL has participated in regional pollution studies to address ozone pollution. New lidar techniques to measure ammonia and fine particulate pollutants are also under development. *ARL* provides advanced regional-scale models such as air quality forecasts and tracking dispersion for emergency management for uses by other agencies. ARL operates the realtime environmental applications and demonstration system (READY) for external users to air quality forecast products.

SECTOR RELEVANCE academia, state and local government

MAIN RESOURCES OFFERED R2, R6, R14

TECHNOLOGY FUNCTIONS T1, T2

CONTACTS

Environmental Technology Laboratory
325 Broadway
Boulder, CO 80303
info@etl.noaa.gov

Air Resources Laboratory
1315 East West Highway
Silver Spring, MD 20910
Tel: 301-713-0684

Department of Defense (DOD)

WEB SITE www.defenselink.mil/

KEY WORDS see individual programs

PURPOSE to provide the military forces needed to deter war and to protect the security of our country.

PROGRAMS

- Defense Installations and Environment Office (DUSDIE)
  - Defense Environmental Network and Information exchange (DENIX) Environmental Security Technology Certification Program (ESTCP)
- Defense Logistics Agency (DLA)
  - DLA Environmental Products
  - Procurement Technical Assistance Centers
- Industrial Ecology Center
  - National Defense Center for Environmental Excellence (NDCEE)
  - TransAction Technology Group (TTG)
- Strategic Environmental Research and Development Program (SERDP)
- U.S. Army Environmental Center (USAEC)
  - Tri-Service Site Characterization and Analysis Penetrometer System (SCAPS)

SUMMARY *DOD* embraces the three military departments — Army, Air Force and Navy, (the Marines are under the Navy) — and 14 specialized defense agencies. The four armed services are subordinate to their military departments. Environmental responsibilities are managed by each service, with general direction from headquarters by the *Deputy Undersecretary for Installations and Environment,* who reports to the Undersecretary of Defense for Acquisition, Technology and Logistics. Access to information on environmental technologies developed or used by the services is provided through the central *Defense Environmental Network and Information eXchange (DENIX).* While direct expenditures on environmental technology are relatively small compared with the total DOD budget, many activities of the services and DOD specialized agencies, such as the *USAEC* and *SERDP,* focus on environmental remediation and pollution prevention. The base closure and conversion program, for instance, has forced DOD to deal with the legacy of hazardous materials, chemicals and unexploded ordinance. Environmental factors enter into the buying of about $60 billion in weapons, equipment and commercial off-the-shelf items. The defense industry is subject to environmental requirements imposed by federal contracting law. Most of the defense industry's applications of environmental technology are not disclosed. The programs with the most active outreach to nonmilitary users are described in the following pages.

MAIN RESOURCES OFFERED see individual programs

TECHNOLOGY FUNCTIONS see individual programs

SECTOR RELEVANCE defense, industry, military

CONTACT For links to all services and DOD organizations, the most useful entry point is the American Forces Information Service at www.defenselink.mil/afis, but DENIX is the quickest path to environmental technology information.

Defense Installations and Environment Office (DUSDIE)

WEB SITE www.acq.osd.mil/ens/
Report available at: www.denix.osd.mil/denix/Public/News/OSD/EQ00/home.html.

KEY WORDS air quality, bioremediation, energy efficiency, fuels, groundwater, hazardous materials, oxidation, soil remediation, waste reduction

PROGRAMS

- Defense Environmental Network and Information Exchange (DENIX)
  - Environmental Security Technology Certification Program (ESTCP)
  - Industrial Ecology Center
  - National Defense Center for Environmental Excellence (NDCEE)
  - TransAction Technology Group (TTG)
- Strategic Environmental Research and Development Program (SERDP)
- U.S. Army Environmental Center (USAEC)
- Tri-Service Site Characterization and Analysis Penetrometer System (SCAPS)

FIGURE 5.10
DOD environmental security programs address climate change, energy efficiency and ozone protection, as well as sustainable land management in addition to compliance with federal and state environmental regulations.

PURPOSE to reduce the total ownership costs of the military infrastructure while providing focused investments to efficiently support war, fighter deployment and training capabilities, and ensure secure, safe and environmentally sound living and working conditions.

SUMMARY *DUSDIE* seeks to reduce the cost of cleanup and compliance, promote technology innovations, obtain better and cheaper environmental performance and support community revitalization with fast-track cleanup of closed or realigned facilities. DOD technology development and transfer programs support not just its military missions, but its far-flung network of bases and installations. Because of the legacy environmental impacts at these facilities, which border on or are in the midst of urban, rural, desert, wetland and other ecological systems, DOD is heavily involved in community and industry partnerships. Site remediation is a major component of DOD programs in connection with its *base realignment and closure* (BRAC) program, which focuses on turning these properties into economic and environmental assets. Although DOD budgets have shrunk since the end of the Cold War, the department still has one of the largest R&D budgets in the federal government, as well as several technology development and transfer programs that relate to all phases of environmental protection and enhancement. Among the hundreds of environmental programs, only a few allow the general public to access their resources. Contractors to DOD and other federal agencies, state and local governments that partner with DOD and some international collaborators can obtain access privileges through programs for which they work or registration menus on the Web.

MAIN RESOURCES OFFERED R1, R2, R3, R4, R5, R6, R8, R10, R11, R12, R13, R14

TECHNOLOGY FUNCTIONS T1, T2, T3, T4, T5, T6, T7, T8, T9, T10, T11

SECTOR RELEVANCE See DENIX

CONTACT

ODUSD (ES)
3400 Defense Pentagon
Room 3E791
Washington, DC 20301-3400

Defense Information Network and Information Exchange (DENIX)

WEB SITE www.denix.osd.mil/denix/Public/ES-Programs/ES/es.html

KEY WORDS climate change, construction waste, ecological risk assessment, ecosystem, energy efficiency, explosives, fire-fighting fuels, invasive species, land use, lead, oil, ozone protection, perchloroethylene, polychlorinated biphenyls, toxic waste, watershed

PURPOSE to provide DOD personnel in environmental security and safety and occupational health with timely access to environmental legislative, compliance, restoration, cleanup and DOD guidance information. It is intended to serve as a central electronic information exchange among environmental professionals worldwide.

SUMMARY *DENIX* provides information, communication and collaboration services to DOD environmental personnel, state regulatory and educational environmental offices, the international environmental community and the general public. The DOD, state and international menus are each distinct extranets for environmental professionals within those communities. User login and password are required for these three areas. The registration page link contains information on eligibility. DOD contractors with an existing contract and DOD point-of-contact are also welcome to participate in the restricted areas. The public menu contains that information which is available to the general public.

MAIN RESOURCE OFFERED: R2

TECHNOLOGY FUNCTIONS T1, T2, T3, T4, T5, T6, T7, T8, T9, T10, T11

SECTOR RELEVANCE agriculture, automotive, aviation, chemicals, construction, defense contracting, fisheries, forestry, land management, manufacturing, marine, petroleum, state and local government

CONTACT

To send an e-mail inquiry:

WEB SITE www.denix.osd.mil/denix/contact.html

Environmental Security Technology Certification Program (ESTCP)

WEB SITE www.estcp.org

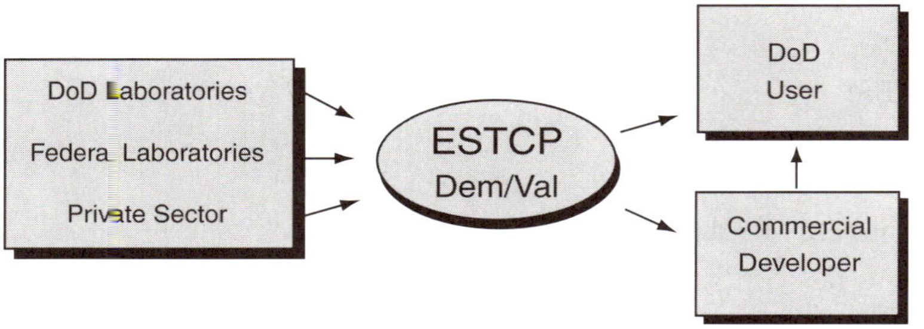

FIGURE 5.11
The Environmental Security Technology Certification Program's (ESTCP) laboratories send technology proposals to ESTCP. After competitive selection, ESTCP supports the demonstration and validation of technologies in the field. ESTCP then encourages implementation and commercialization.

KEY WORDS air sparging, bioremediation, emission, material substitution, monitoring, paints, phytoremediation, recycling, site assessment, VOCs

PURPOSE to demonstrate and validate promising innovative technologies that target DOD's most urgent environmental needs.

SUMMARY *ESTCP* projects address cleanup, compliance, pollution prevention and unexploded ordnance (UXO). They involve partnerships with technology developers, DOD users, independent validators and the regulatory community. Private-sector organizations are solicited via a broad agency announcement issued in the *Commerce Business Daily* in mid-January. Pre-proposals are requested for submittal in March. For those selected to provide proposals, a DOD partner will be chosen to aid in identifying an appropriate demonstration site and in conveying DOD needs.

MAIN RESOURCES OFFERED R1 R3, R6, R10, R13

TECHNOLOGY FUNCTIONS T1, T2, T4, T6-T12

SECTOR RELEVANCE Various

CONTACT

ESTCP Program Office
901 North Stuart Street, Suite 303
Arlington, VA 22203
Tel: 703-696-2117
Fax: 703-696-2114

Industrial Ecology Center (IEC) — U.S. Army (URL blocked since September 11, 2001)

KEY WORDS corrosion, emissions, energy, hazardous waste, materials, solid waste

PURPOSE to reduce the overall throughput of materials and energy, whether they depart as emissions, products, or wastes (hazardous or solid) in the U.S. Army, DOD and the defense contracting community.

PROGRAMS
- Corrosion Prevention and Control (CPC)
- National Defense Center for Environmental Excellence (NDCEE)
- Casting Emissions Reduction Program (CERP)

SUMMARY *IEC* focuses on the application of ecological principles to Army, DOD and related commercial industrial processes. The *CPC* program addresses the serious corrosion concerns for Army weapon systems. DOD's cost for corrosion prevention and control has been estimated at approximately $10 billion per year[6] The program consists of alternate environmentally acceptable technologies, total quality environmental management, environmental and energy cost analysis and ecological management[7] in support of defense acquisition reform issues (green acquisition). The *NDCEE*'s mission is to transition environmentally acceptable materials and processes to defense industrial activities and private industry, provide training that supports the use of new environmentally acceptable technologies and, where appropriate, support R & D to transition new technologies. Private industries initiate projects within the NDCEE Program, which is managed by Concurrent Technologies Corporation and monitored by the IEC. The *CERP* evaluates, qualifies and tests possible replacement materials or manufacturing processes for the reduction of hazardous emissions from metal-casting processes.

MAIN RESOURCES OFFERED R4, R10, R12, R15

TECHNOLOGY FUNCTIONS T2, T3, T4, T5, T6, T7, T8, T9, T11

SECTOR RELEVANCE defense, foundries, industry, manufacturing

CONTACT

Public Affairs
Picattiny Arsenal, NJ 0790890
Tel: 973-724-6364
Fax: 973-724-6582

NDCEE
www.ndcee.ctc.com
Richard Eichholtz
Contracting Officer Representative
Tel: 410-436-5910
E-mail: Dick.Eichholtz@daapgea050.apgea.army.mil

Frederick J. Mulkey
Program Manager
Tel: 814-269-2877
E-mail: mulkey@ctc.com

Strategic Environmental Research and Development Program (SERDP) (*also see Environmental Security Technology Certification Program*)

WEB SITE www.serdp.org/ (or .com)

KEY WORDS air sparging, aquifer, bioavailability, bioremediation, biosorption, CO_2, corrosion, HAPs, heavy metals, pesticide, refrigeration, solvent, surfactant, VOCs

PURPOSE to improve mission readiness through environmental research by accelerating cost-effective cleanup of contaminated defense sites; facilitating full compliance with environmental laws and regulations; enhancing training, testing and operational readiness through prudent conservation measures; and reduce defense industrial waste streams through aggressive pollution prevention.

SUMMARY *SERDP* is the DOD's corporate environmental R&D program, planned and executed in full partnership with the DOE and the EPA, with participation by numerous other federal and non-federal organizations. Within its broad areas of interest, the program focuses on cleanup, unexploded ordnance or UXO, compliance, conservation and pollution prevention technologies. SERDP funds environmental research and development for both government and private sectors, including demonstration test sites. Its national environmental technology test sites (NETTS) program is a Tri-Services and EPA effort in partnership with the private sector to reduce the cost and accelerate the pace of remediation technology development and deployment. The NETTS program facilitates transfer of technology from research to full-scale use and offers test beds at three sites: Dover Air Force Base (Dover, Delaware); McClellan Air Force Base (Sacramento, California); and Naval Construction Battalion Center (Port Hueneme, California). Transferring technology is a principal goal of SERDP. Information resources include quarterly bulletins, annual reports to congress and enviroscience e-print services. In addition, it hosts an annual (in fall) Partners in Environmental Technology Technical Symposium and Workshop.

MAIN RESOURCES OFFERED R1, R3, R4, R5, R14

TECHNOLOGY FUNCTIONS T1, T2, T3, T4, T5, T6, T7, T9, T10, T11, T12

SECTOR RELEVANCE various, including defense contracting, aviation, chemicals, defense contracting, maritime, metals, manufacturing, paint

CONTACT

SERDP Program Office
901 North Stuart Street, Suite 303
Arlington, VA 22203
Tel: 703-696-2117
Fax: 703-696-2114

Sample Technology: Passive Diffusion Membrane Sampler (SERDP)

KEY WORDS groundwater, volatile organic compounds

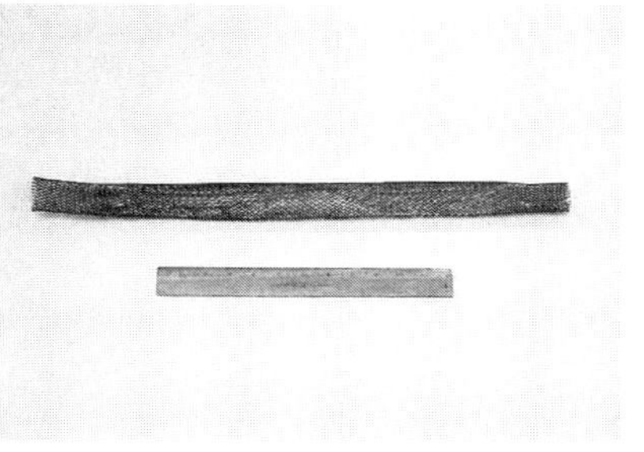

FIGURE 5.12
Diffusion membrane groundwater sample (shown next to 12-in. ruler).

ENVIRONMENTAL CHALLENGE Conventional purge-and-sample methods involve significant labor for purging and other costs, such as disposal of contaminated water brought to the surface.

PERFORMER/DEVELOPER U.S. Geological Survey

RESULTS Diffusion membrane groundwater samplers have demonstrated technical and cost advantages over conventional purge-and-sample methods. Many times, the diffusion sampler data appear to be more representative of the true nature of VOC contamination in the groundwater immediately adjacent to the well screen. Potential savings are estimated between $275 and $600 per sample (the greater amount if purge water must be disposed of off-site). Use of the samplers has been approved by Region IX of EPA and two California regulatory agencies. Several vendors have been granted licensing rights to commercialize the technology. (See further description of deployments under the Interstate Technology and Regulatory Cooperation entry, Chapter 9.)

TECHNOLOGY EMPLOYED A low-density, water-filled polyethylene diffusion membrane collects groundwater samples for VOC analysis. The 1.5-foot-long samplers attached to a weighted rope are lowered into a monitoring well for a minimum of 2 weeks until VOC equilibrium occurs between the well water and deionized water within the samplers.

The U.S. Army Environmental Center (USAEC)

WEB SITE
www.aec.army.mil/usaec/index.html

KEY WORDS aquaculture, coastal, dredging, ecosystem, hydraulics, site remediation, wetlands

FIGURE 5.13
USAEC supports a variety of P2 and environmental compliance programs. (USAEC, with permission.)

PURPOSE to provide environmental compliance support for Army installation restoration and participate in environmental technology demonstration and transfer.

SUMMARY *USAEC*'s support encompasses the four pillars of the Army's environmental strategy for the 21st century: compliance, restoration, prevention and conservation. This environmental support includes conducting site investigations, testing groundwater, reviewing Army environmental requirements and implementing new sampling technologies and providing links to pollution prevention resources, such as recycled-products catalogs. Located at Aberdeen Proving Ground in Edgewood, Maryland, the USAEC works with the environmental staffs of major Army and U.S. Army Corps of Engineers' divisions and districts, field operating activities, installations and facilities nationwide and around the world. Its *Environmental Laboratory* (EL) offers some licensing of patents. The parent *Engineering Research and Development Center (ERDC)* makes available research results on line, including from the EL. The Center's *Fielding Environmental Solutions* (FES) online newsletter provides information on recently published documents, field demonstrations of innovative technologies and technology transfer activities.

MAIN RESOURCES OFFERED R1, R2, R8, R11, R12, R13

TECHNOLOGY FUNCTIONS T1, T2, T3, T5.T6, T7, T8, T10, T11, T12

SECTOR RELEVANCE Various

CONTACTS

USAEC
Attn: SFIM-AEC-PA
5179 Hoadley Road
Aberdeen Proving Ground, MD 21010-5401
Tel: 410-436-2556
Fax: 410-436-1693
E-mail: www.aec.army.mil/usaec/contactus.html

Pollution Prevention & Environmental Technology Division
Hotline: 1-800-USA-3845

ERDC
US Army Engineer Research and Development Center
3909 Halls Ferry Road
Attn: CEERD-PA-Z
Vicksburg, MS 39180-6199

Environmental Laboratory
Ginny Dickerson, CEWES-EV-I (Contract Support)
3909 Halls Ferry Rd.
Vicksburg, MS 39180-6133
Tel: 601-634-4261
E-Mail: dickerv@wes.army.mil

To subscribe to the FES newsletter: send an e-mail message without a subject to listerv@aec.apgea.army.mil, with the message: "subscribe fielding environmental solutions" with your first and last names.

Sample Technology: Tri-Service Site Characterization and Analysis Penetrometer System (SCAPS)

KEY WORDS site assessment, remediation

FIGURE 5.14
Mobile site characterization units have documented savings of over $6 million at DOD and other sites.

SUMMARY The use of SCAPS for detecting petroleum, oils and lubricants in subsurface soil and water typically saves approximately 50% of the cost of site characterization compared with conventional methods. The SCAPS truck utilizes vertical hydraulic rams to force a cone penetrometer into the ground at a speed of 2cm/sec to depths of approximately 25–50m in nominally consolidated fine-grained soils. Data are continuously collected and recorded with 2cm spatial resolution and permits 3-dimensional visualization of subsurface soil stratigraphy and regions of potential contamination. SCAPS has been evaluated and recommended under the EPA Superfund Innovative Technology Program. In its 1998 edition of internationally recognized trade standards, the American Society of Testing and Materials (ASTM) includes procedures for using the laser-induced fluorescence (LIF) sensor, which employs ultraviolet laser energy to induce fluorescence in contaminants. The LIF technology has received regulatory approval from California and the Interstate Technology Regulatory Cooperation Workgroup.

The U.S. Army Engineer Waterways Experiment Station (WES) under the sponsorship of the U.S. Army Environmental Center (AEC) initiated the development of SCAPS. Eleven patents have been granted and four are pending. SCAPS technologies have been transitioned to the private sector via licensing agreements, cooperative research and development agreements and technology reinvestment programs.

CONTACT

John Ballard
USAE Waterways Experiment Station
Attn: CEWES-EP-J, 3909
Halls Ferry Road,
Vicksburg, MS 39180-6199,
Tel: 601-634-2446
Fax: 601-634-2732

Department of Energy (DOE)

WEB SITE www.energy.gov/

> "Energy which is environmentally safe in its delivery is no longer something … that we can take for granted."—**Energy Secretary Spencer Abraham**

KEY WORDS biodegradation rate, bioremediation, climate change, energy efficiency, environmental assessment, hazardous wastes, indoor air pollution, industrial technology, long-term monitoring, renewable energy, site remediation

PURPOSE to foster a secure and reliable energy system that is environmentally and economically sustainable, to be a responsible steward of the nation's nuclear weapons, to clean up our own facilities and to support continued United States leadership in science and technology.

PROGRAMS

- Energy Efficiency and Renewable Energy (EERE)
  - Building Technologies, Weatherization and Intergovernmental
  - Office of Industrial Technologies (OIT)
- Energy Information Administration (EIA)
- National Laboratories
  - Lawrence Berkeley National Laboratory (LBNL)
  - National Energy Technology Laboratory (NETL)
  - Oak Ridge National Laboratory (ORNL)
- Office of Scientific and Technical Program Information
- Environmental Management Science & Technology Program (EMSTP)

SUMMARY Although the bulk of the Office of Science and Technology's (OST) budget is defense-related, or focused on fossil fuel and nuclear energy, achieving renewable energy and energy efficiency continues to be a major goal through the *EERE* program. Many other elements of DOE support *EERE*'s work. *EIA*, for example, is the government's central repository of energy supply-and-demand data. It conducts special studies on sectors and industries, including the environmental impacts of energy use. The *National Laboratories* have taken on projects related to *EERE*. Those listed above have extensive programs or projects to assist in the development and dissemination of environmental technologies. The *OST* disseminates scientific and technical information related to environmental technologies and administers DOE's patent licensing.

MAIN RESOURCES OFFERED R1, R2, R3, R4, R5, R6, R7, R8, R9, R10, R11, R12, R13

TECHNOLOGY FUNCTIONS T1, T2, T3, T4, T5, T6, T7, T8, T9, T10, T11

SECTOR RELEVANCE nuclear power, utilities, manufacturing, construction, heating and cooling, HVAC, state and local government, transportation

CONTACT

U.S. Department of Energy
1000 Independence Ave., SW
Washington, DC 20585
Tel: 1-800-dial-DOE
www.energy.gov/contacts/index.html

Office of Energy Efficiency and Renewable Energy (EERE)

WEB SITE www.eren.doe.gov/

FIGURE 5.15
DOE clean energy programs focus on several technology areas (DOE, with permission).

KEY WORDS brownfields, building material, energy efficiency, fuels, geothermal, hydrogen, photovoltaics, water and space heating

PROGRAMS

- Building Technologies, Weatherization and Intergovernmental
- Industrial Technologies

PURPOSE to strengthen America's energy security, environmental quality, and economic vitality in public-private partnerships that enhance energy efficiency and productivity; bring clean, reliable and affordable energy technologies to the marketplace; and enhance (Americans') energy choices and their quality of life.

SUMMARY In July 2002, *EERE* reorganized to strengthen its focus on programs and these partnerships. Instead of just five energy sectors—industry, transportation, buildings, power, and Federal agencies, EERE is now organized around 11 energy programs:
Biomass, Building Technologies, Distributed Energy & Electricity Reliability, Federal Energy Management, FreedomCAR and Vehicle Technologies, Geothermal Technologies, Hydrogen, Fuel Cells & Infrastructure Technologies, Industrial Technologies, Solar Energy Technology, Wind & Hydropower Technologies, Weatherization and Intergovernmental. Regional offices work with state and local governments, regional industries and other stakeholders to promote adoption of EERE technologies. EERE provides financing, technology development and other assistance through a number of focused programs such as bioenergy, renewable energy for redevelopment of Brownfields, co-generation, schools and solar home heating.

MAIN RESOURCES OFFERED R1, R2, R3, R4, R5, R6, R7, R8, R9, R10, R11, R12, R13, R14

TECHNOLOGY FUNCTIONS T1, T2, T3, T4, T5, T6, T7, T8, T9, T10, T11

SECTOR RELEVANCE agriculture, construction (residential and industrial), heating and cooling, manufacturing, state and local government, transportation, utilities, water treatment

CONTACT

EERE
U.S. Department of Energy
Mail Stop EE-1
Washington, DC 20585
Tel: 202-586-9220
www.eren.doe.gov/eere/contacts.html

Building Technologies, Weatherization and Intergovernmental

WEB SITES www.eren.doe.gov/building.html and www.eren.doe.gov/weatherization.html

KEY WORDS building materials, desiccant, energy efficiency, heating and cooling, insulation, lighting, water

PURPOSE in partnership with industry and government, to develop, promote and integrate energy technologies and practices to make buildings more efficient and affordable and communities more livable.

PROGRAMS (Specialized information)

- Homeowners
- Commercial Buildings
- Builders and Designers
- Researchers
- Partners and Public Officials

FIGURE 5.16
DOE targets exterior and interior features of buildings to save energy. (Lawrence Berkeley National Laboratory, with permission).

SUMMARY *BTS* activities include research and development; increasing the minimum efficiency of buildings and equipment through codes, standards and guidelines; and encouraging use of energy efficiency and renewable energy technologies and practices through technology transfer and financial assistance. Many of the new-building technologies mentioned on the site come from one or more of DOE's laboratories. (See, for example, Lawrence Berkeley and National Renewable Energy lab sites.) This site provides a wealth of data, however, on energy consumption and potential energy savings by type of building, state and other categories. It also links to the DOE/EPA ENERGY STAR® site (see description under EPA). According to BTS, ENERGY STAR® equipment can save about 40% on the typical U.S. household's $1300 annual energy bill.

MAIN RESOURCES OFFERED R1, R3, R6, R7, R10, R11, R12

TECHNOLOGY FUNCTIONS T1, T4, T5, T6, T7

SECTOR RELEVANCE appliance manufacturing, architect and engineering firms, building management, construction materials, motor vehicles, native American tribes, property management, real estate, state and local government, utilities.

CONTACT

EERE
U.S. Department of Energy
1000 Independence Ave., SW
Washington, DC 20585
Hotline: 1-800-dial-DOE
www.eren.doe.gov/organization/contacts.html

Industrial Technologies (OIT)

WEB SITE www.oit.doe.gov/

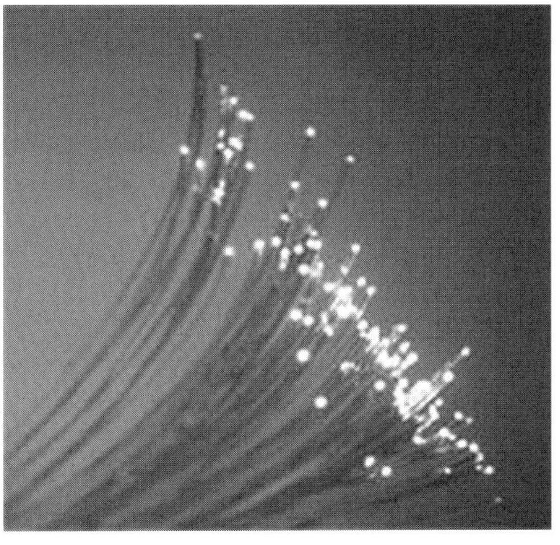

FIGURES 5.17 AND 5.18
Glass and forest products are two of nearly a dozen industries targeted by DOE for improving energy efficiency.

KEY WORDS ceramics, combustion, composites, compressed natural gas, energy efficiency, materials

PURPOSE to work in partnership with U.S. industry to develop and deliver advanced technologies that increase energy efficiency, improve environmental performance, and boost productivity

PROGRAMS

- Best Practices
- Enabling Technologies
- Financial Assistance
- Industries of the Future
- International
- National Industrial Competitiveness through Energy, Environment and Economics (NICE3)
- State Industries of the Future

SUMMARY *Best Practices* offers in-plant assessments, tools, training and other resources to companies to identify energy efficiencies and cleaner, more productive technologies. *Enabling Technologies* offers R&D cost sharing in developing combustion, sensor and materials technologies. *Financial Assistance* provides support to inventors and small businesses to launch energy-saving inventions and innovations. *Industries of the Future* provides cost sharing to collaborative R&D partnerships in nine energy-intensive industries (see Sector Relevance below). *International* promotes U.S. business leader dialogues with other nations on energy efficiency and sustainability. *NICE3* promotes demonstrations of energy efficient technologies through state–industry partnerships. *State Industries of the Future* extends the benefits of the Industries of the Future strategy to the state and regional level. A quarterly newsletter, *OIT Times*, is available online.

RESULTS Cumulative energy savings from more than 140 completed and tracked OIT projects and other programs is approximately 1.6 quadrillion Btu, representing production cost savings of $6.5 billion. Environmental benefits include keeping 121 million tons of CO_2, 246,000 tons NO_x, 463,000 tons SO_x and 124,000 tons particulates out of the air (see Summary of Program Results Impacts at www.pnl.gov/impacts). In addition to energy savings, productivity gains from these technologies are also considerable. An average of about $55,000 in potential annual savings for each manufacturer have been identified.

MAIN RESOURCES OFFERED R1, R3, R4, R5, R6, R7, R8, R9, R10, R11, R12, R13, R14

TECHNOLOGY FUNCTIONS T1, T2, T4, T5, T6, T7, T8, T10, T11

SECTOR RELEVANCE agriculture, aluminum, chemicals, forest products, glass, metal casting, mining, petroleum, state and local government and steel. (Note: other industries may be added in the future and other OIT programs serve most industries.)

CONTACT

Tel: 202-586-9232

Hotline: 1-800-862-2086

E-mail: clearinghouse@ee.doe.gov

Sample Resource

The OIT Clearinghouse

www.oit.doe.gov/clearinghouse

The OIT Clearinghouse offers online responses to technical questions on energy and environmental technologies from clearinghouse engineers, other technical staff and customer service specialists. Typical queries regard such matters as justifying preventive maintenance for a steam system, providing cost–benefit information on use of higher efficiency motors, finding grants for demonstrating an energy-efficient process, and energy "roadmaps" for resource-intensive industries.

USER COMMENTS "The real world experience of your organization places you in a unique position to provide solid advice and guidance. Keep up the good work."

CONTACTS

Tel: 1-800-862-2086 (9 am–8pm, EST, Monday-Friday)
Fax: 360-586-8303
E-mail: clearinghouse@ee.doe.gov

National Industrial Competitiveness through Energy, Environment and Economics (NICE3)

WEB SITE www.oit.doe.gov/nice3/

KEY WORDS energy efficiency

PURPOSE to promote energy efficiency, clean production and economic competitiveness in industry.

SUMMARY *NICE3* provides one-time project grants (up to $525,000) to state and industry partnerships (large and small business) to develop and demonstrate advances in energy efficiency and clean production technologies. Non-federal cost share must be at least 50% of the total cost of the project. In total, *NICE3* has sponsored 91 projects, with more than half going to small businesses. *NICE3* has leveraged $26.3 million in federal funds, with $81.8 million in state and industry funds since 1991. (Note: no new project funding is available for fiscal year 2002 ending Sept. 30, 2002.) Sample project histories in the OIT Industries of the Future program are online.

MAIN RESOURCES OFFERED R1, R3, R6, R9, R12, R13

TECHNOLOGY FUNCTIONS T1, T2, T4, T6, T7, T8, T9, T10, T11, T12

SECTOR RELEVANCE agriculture, aluminum, chemicals, electronics, food, forest products, general manufacturing, glass, metal casting, mining, petroleum, printing, state government, steel and textiles

CONTACT

Lisa Barnett
National Program Manager
NICE3
E-mail: Lisa.Barnett@ee.doe.gov
For financial assistance information, see NICE3 guidelines and OIT contacts on the Web site.

Sample Technology. Robotics Inspection System for Aboveground Storage Tanks

WEB www.oit.doe.gov/factsheets/nice3/pdfs/robotics.pdf

DEVELOPER Don Hartsell Solex (713-963-8600) drh@solexrobotics.com

KEY WORDS carbon dioxide, chemicals, fuels, groundwater, robotics, volatile organic compounds

PROJECT OBJECTIVE: to demonstrate and commercialize a robotic inspection vehicle that operates in aboveground storage tanks filled with liquid. DOE provided Solex a $425,000 grant to test and refine the robot and obtain the required safety certifications. The Texas Natural Resources Conservation Commission managed the grant.

FIGURE 5.19
"Maverick" robot inspects for aboveground tank flaws. (Solarex, with permission)

ENVIRONMENTAL CHALLENGE to detect storage tank leaks early to prevent groundwater and soil pollution. Billions of gallons of gasoline and jet fuel are stored in huge tanks in the petroleum, chemicals and forest products industries around the world. If these tanks leak, the poisonous chemicals they contain can seep into the groundwater. Inspecting the tanks manually for structural integrity and corrosion damage is a tedious, potentially dangerous and expensive task, requiring venting, emptying and cleaning the tank. Increasingly, governments are mandating inspections and the petroleum industry has adopted a voluntary standard for criteria and timing of inspecting aboveground storage tanks.

PROJECT RESULTS Maverick is a certified explosion-proof inspection system for in-service aboveground fluid storage tanks. It reduces the costs of inspections by $50,000 to $500,000 per tank each time it is used. Although there are no requirements for aboveground tank inspections, the use of robots makes it possible to conduct them without taking the tank out of service. Using the robot (1) avoids tank draining, cleaning and downtime; (2) limits exposure of personnel to noxious fumes, hazardous chemicals and confined spaces; (3) reduces CO_2 and VOC emissions and hazardous waste by eliminating the need to vent tanks before cleaning and to dispose of cleaning agents and sludge from tank bottoms; and (4) achieves energy savings of more than 95% per inspection. This calculation takes into account various factors, including avoidance of releasing vapors (Texas, for example, requires burning off the fumes and California has a collection requirement) and the energy wasted from discarding tank sludge contents.

TECHNOLOGY DESCRIPTION designed by researchers at DOE's Idaho National Engineering and Environmental Laboratory and Solex Robotics Systems, Maverick is a remotely operated robot about the size of a suitcase that can be lowered through portals on top of the tank. Connected via a cable to computers in a nearby trailer, the robot travels around the tank floor in the tank liquid. Sensors and imaging systems check the tank bottom for evidence of thinning or leaking, using a multichannel ultrasonic sensor array, onboard video system and sonar positioning.

COMMENTS

> "The in-service inspection was completed safely and in a cost effective manner."—**Mark Sesselman, Amoco**

"The demonstration has led to commercialization and it is being used here and overseas. Inspecting tanks is very dangerous and very time consuming. The program (NICE3) basically allows the innovation to get off the ground. It takes a lot of the fear out of the developer to have a partnership with the state and with DOE. We are watching over their shoulder making sure they are doing it right. The results are reviewed by a whole board of individuals, so there is verification that it works and meets the goals of the proposal."—**Jeff Voorhes, project manager, Texas Natural Resource Conservation Commission**

"The grant and partnerships made it possible for a small business to obtain the interest of companies like Exxon and Shell. Since then, we have had a variety of opportunities presented to us and it has led to the development of additional technology."—**Don Hartsell, president of Solex Robotics**

CONTACTS

Solex Environmental Systems, Inc.
P.O. Box 460242
Houston, TX 77056
Tel: 713-963-8600
E-mail: drh@solexrobotics.com

DOE/NICE3
Lisa Barnett, Program Manager
Tel: 202-586-2012
E-mail: lisa.barnett@ee.doe.gov

Lawrence Berkeley National Laboratory (LBNL)

WEB SITE www.lbl.gov/

KEY WORDS bioavailability, bioremediation, building materials, carbon dioxide (CO_2), climate change, energy efficiency, fuels, heating and cooling, indoor air quality, information technology, lighting, pollution, risk assessment

PURPOSE to conduct research in advanced materials, life sciences, energy efficiency, detectors and accelerators, that serves America's needs in technology and the environment.

PROGRAMS

- Environmental Energy Technologies Division
- Technology Transfer Office
- Center for Environmental Biotechnology (CEB)

FIGURE 5.20
Lawrence Berkeley National Laboratory (LNBL). From microbiology to space, LNBL science contributes to DOE energy and environmental solutions through interdisciplinary R&D. (LNBL, with permission.)

SUMMARY The *Environmental Energy Technologies Division* works in such fields as advanced energy-efficient buildings, software and standards; global climate change; indoor-air quality; environmental diagnostics and environmental health risk assessment. The lab is also working at the cutting edge of research on energy storage and conversion (batteries and fuel cells), processes to reduce the environmental impacts of energy technologies (such as air pollutants) and biological methods of environmental remediation. The *Technology Transfer Office* assists industry to pinpoint research areas of common interest, negotiate rights to the LBNL's intellectual property and discuss current patent and copyright licensing opportunities. The *Center for Environmental Biotechnology* conducts integrated interdisciplinary research in biogeochemical transformations, environmental diagnostics and environmental health risk assessment.

MAIN RESOURCES OFFERED R1, R2, R3, R4, R5, R8, R10, R11, R12, R14

TECHNOLOGY FUNCTIONS T1, T2, T3, T4, T5, T6, T7, T8, T9, T10, T11, T12

SECTOR RELEVANCE architect and engineering firms, agriculture, commercial building management, construction, building maintenance, home builders, HVAC, office equipment, power generation, property management, utilities

CONTACTS

Environmental Energy Technologies Division
E-mail: EETDinfo@lbl.gov

Partnerships, technical assistance

www.lbl.gov/Tech-Transfer/

E-mail: TTD@lbl.gov

Research programs or publications

E-mail: EETDinfo@lbl.gov

Sample Resource: Licenses

www.lbl.gov/Tech-Transfer/collaboration/techs/index.html

Licenses are available in these categories: batteries, biotechnology and medicine, chemical and manufacturing processes, computing, energy saving and efficiency technologies, energy and environment, ion sources and beam technologies, materials science and sensors and controls/instrumentation.

Sample Technology: Aerosol-Based Duct Sealing Technology

www.lbl.gov/Tech-Transfer/success_stories/spinoffs/index.html

LICENSEE Aeroseal, Inc. (Austin, TX)
www.aeroseal.com/

DEVELOPER Lawrence Berkeley National Laboratory

ENVIRONMENTAL CHALLENGE to reduce energy losses through HVAC duct leaks. Each year, residential and commercial duct leakage costs consumers some $5 billion—a loss equal to the annual energy consumption of 13 million cars. A typical house with ducts located in the attic or crawl space wastes about 20% of heating and cooling energy through duct leaks and draws approximately 1 kW more electricity during peak cooling periods. Studies indicate that sealing leaks could save close to one quadrillion Btus per year.[8]

RESULTS Aerosol sealing of duct leaks reduces heating and cooling energy use by up to 30%. Compared with other methods of sealing ducts in an existing structure, the application is minimally disruptive to the building and its occupants.[9]

DESCRIPTION Designed for use in residential and small commercial building heating and cooling ducts, the duct sealer blows aerosolized adhesive particles into the duct system, depositing them only at the leakage sites, thus sealing the leaks.

National Energy Technology Laboratory (NETL)

WEB SITE
www.netl.doe.gov/

KEY WORDS climate change, energy efficiency, fossil fuels, metals, oil, site remediation, watersheds

PURPOSE to assure that U.S. fossil energy resources can meet increasing demand for affordable energy without compromising the quality of life for future generations of Americans.

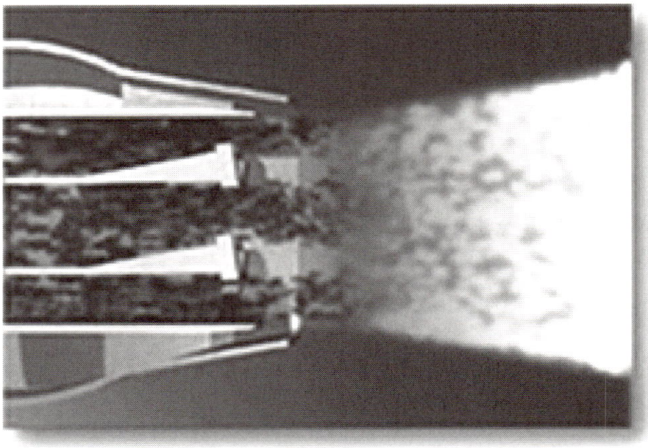

FIGURE 5.21
Low NOx coal combustion systems like this developed with industry and co-funded by the Energy Department hold promise of containing costs of electricity production while meeting more stringent nitrogen oxide emission standards.

SUMMARY *NETL* carries out a number of DOE industry and university programs involving technology development and demonstration. Along with advanced technologies related to fossil fuels—coal, natural gas and oil—the lab has broadened its mission to support development and deployment of environmental technologies, such as carbon sequestration. Research areas include capture and storage, geologic, ocean and terrestrial sequestration, advanced CO_2 conversion and reuse and modeling and analysis. CO_2 conversion processes, occurring in nature, are being investigated as a means to both reduce net carbon emissions and provide significant collateral benefits.

MAIN RESOURCES OFFERED R1, R2, R4, R6, R8, R10, R11, R13, R14

TECHNOLOGY FUNCTIONS T1, T2, T4, T6, T8, T9, T10, T11

SECTOR RELEVANCE fossil fuels, industrial boilers, metal fabrication, oil and gas drilling, power generation,

CONTACT

Regional Public Affairs Manager
National Energy Technology Laboratory
Tel: 412-386-4645 or 412-337-8553
E-mail: anna@netl.doe.gov

Oak Ridge National Laboratory (ORNL)/DOE

WEB SITE www.ornl.gov/

FIGURE 5.22
DOE's Oak Ridge lab investigates a range of energy and environment technologies. (ORNL, with permission.)

KEY WORDS energy efficiency

PROGRAMS

- The Buildings Technology Center (BTC)
- The Center for Transportation Analysis (CTA)
- Energy Efficiency and Renewable Energy Program (EERE)

PURPOSE to conduct basic and applied research and development in key areas of science; increase the availability of clean, abundant energy; restore and protect the environment; and contribute to national security.

SUMMARY *ORNL* pioneers the development of new energy sources, technologies and materials and the advancement of knowledge in the multiple scientific disciplines. In the area of environmental management, the lab supports the cleanup of DOE's environmental legacy, principally at DOE sites, but with increasing emphasis on technology transfer. *BTC* facilities are available to manufacturers, universities and other organizations for proprietary and nonproprietary R&D. *CTA* conducts R&D in transportation energy and environmental issues. Examples of ORNL partnerships include Secat, a University of Kentucky-affiliated, for-profit company that serves as a vehicle to provide economic, environmental and energy benefits to the aluminum industry. *EERE* performs R&D, often in partnership with states, on clean and efficient energy for buildings, industries, power generation and transportation. An example of its environmental data repositories is the world-ecosystem-complex dataset for interpreting the role of vegetation in global GHG cycling.

MAIN RESOURCES OFFERED R2, R5, R8, R10, R13, R14

TECHNOLOGY FUNCTIONS T4, T5, T7, T8, T10

SECTOR RELEVANCE aluminum, commercial buildings, motor vehicles, power generation, state government, transportation

CONTACT

For Technology Transfer and Economic Development Directorate:
Oak Ridge National Laboratory
P.O. Box 2008
Oak Ridge, TN 37831
Tel: 865-241-7613
www.ornl.gov/directorates/tted_org.html

Office of Scientific and Technical Information (OSTI)

WEB SITE www.osti.gov/

KEY WORDS Various

PURPOSE to disseminate scientific and technical information from DOE R&D, including in environmental sciences.

SUMMARY *OSTI* draws on information from its laboratories, 7000 research organizations, foreign countries and other agencies. Its *EnergyFiles: The Virtual Library of Energy Science and Technology* includes the *DOE Information Bridge* for full-text OSTI reports, *PubSCIENCE* for scientific journal literature and the *PrePRINT Network*, which provides a searchable gateway to preprint servers. The *Bridge*, for example, is fully searchable by such parameters as title, author, publication date, sponsor, etc., and can produce bibliographies, counts and full text. A search for underground storage tanks produced 843 matches, including "specialized video systems for use in underground storage tanks." OSTI also administers the DOE patent licensing program (see sample resource).

MAIN RESOURCE OFFERED R2

TECHNOLOGY FUNCTIONS T1, T2, T3, T4, T5, T6, T7, T8, T9, T10, T11, T12, T13

SECTOR RELEVANCE Various

CONTACT

OSTI
P.O. Box 62
Oak Ridge, TN 37831
Tel: 865-576-1188 (general information)
865-576-8401(documents)
Fax: 865-576-2865
E-mail: OSTIWebmaster@osti.gov
www.osti.gov/osticomment-form.html

Sample Resource: DOE Patent Licensing (OSTI)

WEB SITE www.osti.gov/gencoun/gchome.html

DOE owns approximately 1500 unexpired U.S. patented inventions and, in some cases, foreign counterparts. Generally, these are available for license to applicants with satisfactory plans for their commercial use. Exclusive licenses require a determination, after public notice and opportunity for comment, that the invention is not likely to be commercialized on a nonexclusive basis. Technical assistance from the laboratory where the invention arose may be available. In addition, various laboratories (see, for example, Lawrence Berkeley Laboratory) own and directly license a significant number of inventions. Some contractor-owned patents resulting from DOE sponsorship may also be available for licensing. Two patent databases are currently available: the *Current Release* database, recommended for the frequent user, and the *Cumulative Database*, covering technologies developed by DOE laboratories or by DOE-contracted researchers since 1978.

SAMPLE PATENTS ***(note: these have been listed, but are not necessarily still available)***

Current Release Patent Database. Solar reduction of CO_2. The red shift of the absorption spectrum of CO_2 with increasing temperature, along with CO stabilization, are the foundation of a process for direct solar reduction of CO_2 to liquid fuels. The result is a process capable of using renewable solar energy to reduce CO_2 directly from the atmosphere or stationary sources for recycle as a liquid fuel resulting in a net decrease in CO_2 in the atmosphere and decreased demand on fossil fuel. The process includes in sequence: preparation of adequately pure CO_2 feed; preheating CO_2 to near 1800 K; breaking the CO_2 bonds to produce CO, O_2 and O; quenching the hot product mix by rapid admixture with room-temperature CO_2; removing excess thermal energy to produce electricity, heat new CO_2 and enable subsequent process steps; removing the remaining CO_2 from the gas; removing oxygen from the gas; converting part of the product CO to H_2 by the shift reaction; and, finally, catalytically synthesizing methanol from the synthetic gas.

DOE Cumulative Patent Database. Method for formation of subsurface barriers using viscous colloids. A method is described for formation of subsurface barriers using viscous liquids where a viscous liquid solidifies at a controlled rate after injection into soil and forms impermeable isolation of the material enclosed within the subsurface barriers. The viscous liquid is selected from the group consisting of polybutenes, polysiloxanes, colloidal silica and modified colloidal silica of which solidification is controlled by gelling, cooling or cross-linking. Solidification timing is controlled by dilution, addition of brines, coating with alumina, stabilization with various agents and by temperature.

CONTACT

Robert J. Marchick
GC-62, 1000 Independence Avenue, SW,
Washington, DC 20585
Tel: 202-586-4792
Fax: 202-586-2805
E-mail: robert.marchick@hq.doe.gov

The Office of Environmental Management, Characterization, Monitoring & Sensor Technology Crosscutting Program (CMST-CP)

WEB SITE www.cmst.org

KEY WORDS contaminant, environmental assessment, hazardous waste, site remediation

PURPOSE to apply unique expertise and innovative solutions for environmental characterization and monitoring problems to help accomplish safe and cost-effective closure of DOE sites.

SUMMARY *CMST-CP* supports DOE-EM characterization and monitoring R&D revolving around subsurface contaminants and tanks, transuranic and mixed waste, deactivation and decommissioning and nuclear materials. The CMST Web site contains a wealth of information on innovative technologies developed with DOE-EM funding as well as links to five focus area homepages and to other DOE-EM resources. Funding for CMST, however, has been eliminated in the fiscal 2003 budget (beginning October 1, 2002, so future development of technologies will be halted at that time. The Reports and Publications link has progress reports, published quarterly, program reports for the focus areas, and Rainbow Books, containing technology summaries for past fiscal years (these have been discontinued). The May/June 2001 issue of *RadWaste Solutions,* has an article describing CMST programs.

MAIN RESOURCES OFFERED R1, R2, R4, R5, R6, R10, R12, R13

TECHNOLOGY FUNCTIONS T1, T10, T11

SECTOR RELEVANCE federal government, state and local governments, various industries

CONTACT

Beth Moore
US DOE, Office of Science and Technology
Forrestal Bldg, Room 3E-066
Washington, DC 20585
Tel: 202-586-6334
Fax: 202-586-1492
E-mail: beth.moore@em.doe.gov

Endnotes

[1]EPA, Office of Water Fact Sheet, February 2000.

[2]See, for example, Environmental Defense Fund, Murky Waters: the environmental effects of aquaculture in the United States. For a copy, **Contact:** Becky Goldberg, Environmental Defense Fund, phone: (212) 505-2100.

[3]Table 2.1 Energy Consumption by End-Use Sector, 1949–1999, Annual Energy Review, Energy Information Administration.

[4]Indoor Environments Division, Office of Radiation and Indoor Air, EPA, www.epa.gov/air/indoorair/indexold.

[5]BEES measures environmental performance using the environmental life-cycle assessment (LCA) approach specified in the ISO 14040 series of standards. All stages in the life of a product are analyzed: raw-material acquisition, manufacture, transportation, installation, use, and recycling and waste management. Economic performance is measured using the ASTM standard life-cycle cost method, which covers the costs of initial investment, replacement, operation, maintenance and repair, and disposal. Environmental and economic performance is combined into an overall performance measure using the ASTM standard for multi-attribute decision analysis. Source: BEES Please brochure. Available: www.bfrl.nist.gov/oae or call (301) 975-6133.

[6]"Corrosion in DOD Systems: Data Collection and Analysis (Phase I)," Harold Mindlin et al., MIAC Report 8, February 1996.

[7]Ecological management is a systems approach to solving environmental problems, incorporating materials and energy flows, processes, specification document review, individuals, health risk and life-cycle costs (definition used by IEC).

[8]Source: Aeroseal, Inc.

[9]Lawrence Berkeley National Laboratory.

6

Environmental Protection Agency (EPA)

EPA

WEB SITE www.epa.gov

KEY WORDS air quality, chemical engineering, chlorine, energy efficiency, greenhouse gases, hazardous chemicals, hazardous waste, indoor heating, soil remediation, solid waste, traffic congestion, watershed assessment, water quality, waste streams

FIGURE 6.1
Vehicle emissions, while reduced, are the focus of new EPA standards coming into force (EPA, with permission).

PROGRAMS

Independent Programs

- Energy Star (EPA/DOE)
- Environmentally Preferable Purchasing (EPP)
- Environmental Technology Commercialization Center (ETC2)
- Small Business Gateway

Regular Programs

- Office of the Chief Financial Officer
- Environmental Finance Program (EFP)
- Office of Policy, Economics and Innovation
    - Sustainable Industry Development Program
    - Community-based Environmental Protection (CBEP)
- Office of Pollution Prevention and Toxics
- Enviroene
- Design for the Environment (DfE)
- Green Chemistry
- Green Engineering (GE)
- Office of Research and Development
- Environmental Monitoring for Public Access and Community Tracking (EMPACT)

- Environmental Technology Verification Program (ETV)
- Environmental Technology Evaluation Center (EvTEC)
  - Sample Technology: Silt Fence Machine
  - National Exposure Research Laboratory
  - National Risk Management Research Laboratory
  - Superfund Innovative Technology Evaluation (SITE)
- Office of Solid Waste and Emergency Response (OSWER)
  - Chemical Emergency Preparedness and Prevention Office (CEPPO)
  - Technology Innovation Office (TIO)
  - Underground Storage Tanks (UST)
- WasteWise
- Office of Wastewater Management (OWM)
- Office of Wetlands, Oceans and Watersheds (OWOW)
- Office of Transportation and Air Quality
- Transportation Air Quality Center (TRAQ)

PURPOSE to protect human health and to safeguard the natural environment (air, water and land).

SUMMARY The EPA promulgates environmental regulations, enforces them and works with industry, state, local and nongovernmental bodies to disseminate information and best practices. It also provides grants and incentives to achieve compliance and pollution-prevention objectives and supports research to develop innovative environmental technologies, measuring tools and management systems. It is frequently the leading agency for U.S. participation in international environmental treaties and organizations. For details on activities, see the summaries under each program entry.

MAIN RESOURCES OFFERED R1, R2, R3, R4, R5, R6, R7, R8, R9, R10, R11, R12, R13, R14

TECHNOLOGY FUNCTIONS T1, T2, T3, T4, T5, T6, T7, T8, T9, T10, T11, T12

SECTOR RELEVANCE Various

CONTACT

Environmental Protection Agency
Ariel Rios Building
1200 Pennsylvania Avenue N.W.
Washington, DC 20460
Tel: 202-260-2090

Energy Star

WEB SITE www.energystar.gov/default.shtml

KEY WORDS air quality, carbon dioxide, energy efficiency

PURPOSE to identify and promote energy-efficient products that reduce carbon dioxide emissions, thus protecting the environment and saving money for consumers and other users.

FIGURE 6.2
Whether for small appliances or new homes, the label signifies energy and dollar savings (Energy Star, with permission).

SUMMARY *Energy Star*, a voluntary labeling program jointly managed by EPA and DOE, provides information on saving energy and money to large and small businesses, homeowners, governments, schools and other institutions. Manufacturing and distributing companies can apply to partner with the government. Once approved, a partner can attach the Energy Star label to its product(s). Since its formation in 1992, the program has expanded to include information and labeling of new homes, most buildings, sector products, residential heating and cooling equipment, major appliances, office equipment, lighting, consumer electronics and more. The Energy Star Web site includes results of product testing, information on the voluntary standards, energy-saving tips, a newsletter and a map showing the location of Energy Star product carriers.

MAIN RESOURCES OFFERED R1, R4, R10, R11, R14, R15 (economic assessment)

TECHNOLOGY FUNCTIONS T2, T4, T6, T7

SECTOR RELEVANCE appliances, building products, computers, heating and cooling, home construction, lighting

CONTACTS

1-888-STAR-YES (1-888-782-7937)
For partnership agreement form:
Energy Star
c/o D&R International
1300 Spring Street, Suite 500
Silver Spring, MD 20910
E-mail: partnership@energystar.gov

Environmentally Preferable Purchasing (EPP)

WEB SITE www.epa.gov/opptintr/epp/

KEY WORDS biodegradable, ozone protection, plastics, recycle/reuse, waste minimization

PURPOSE to encourage and assist U.S. government agencies in the purchasing of environmentally preferable products and services.

SUMMARY The *Environmentally Preferable Purchasing* program has four goals: 1) improving the ability of business and all levels of government to meet existing environmental goals; 2) improving worker safety and health; 3) reducing liabilities incurred by businesses and government for not meeting environmental standards; and 4) reducing safety and disposal costs. According to Executive Order 13101, the definition of *environmentally preferable* is, "… products or services that have a lesser or reduced effect on human health and the environment when compared with competing products or services that serve the same purpose …." The "tools and resources" part of the EPP Web site provides a number of databases or tips for switching to environmentally sound products. Under the purchasing tool header are links to sites that help the reader learn about, among other things, environmentally safe cleaning products, as well as how to hold greener conferences and meetings.

There is also a link to a database of product-specific information. The database leads the reader to six different organizations that display vendor lists for environmentally safe supplies. These organizations are: 1) The U.S Environmental Protection Agency *Comprehensive Procurement Guidelines*, with eight categories of EPA products designated safe; 2) the EPA *Energy Star* program, with products from all aspects of consumer life; 3) the Canadian *Environmental Choice* program, with products ranging from paints to plastics to appliances; 4) Germany's *Blue Angel* program, which has dozens of specific categories of environmentally safe products; and 5) the *Nordic Swan* program, with a list of product groupings that lead to complex tables of individual products and phone numbers to reach the manufacturers.

The EPP site also offers links to documents related to the EPP and links to other programs useful for creating an environmentally sound business. One of these programs is the *Federal Trade Commission's Green Guides,* which helps advise advertisers on use of environmental claims in advertising.

MAIN RESOURCES OFFERED R1, R2, R15

TECHNOLOGY FUNCTIONS T4, T6, T8, T13

SECTOR RELEVANCE advertising, government contracting, government vendors

CONTACT

EPP Program: epp.pilot@epa.gov

There is also a group e-mail list at www.epa.gov/opptintr/epp/gemail.htm

Environmental Technology Commercialization Center (ETC2)

WEB SITE www.etc2.org/index.htm_

KEY WORDS Various

PURPOSE to enhance EPA's ability to provide less costly, cutting-edge environmental technologies to private companies.

SUMMARY The *Environmental Technology Commercialization Center*, run by Battelle, transfers technology from the government to private industries. ETC2 locates technology developed within EPA laboratory facilities that meets the needs of private companies. ETC2 assists businesses to develop commercialization strategies and perform market opportunity analyses. It also conducts technology dialogues and symposia with technical firms and facilitates joint research and development between industry, universities and federal labs through CRADAs (Cooperative Research and Development Agreement). Resources available from ETC2 online include case studies, descriptions of technology available for licensing, and a brochure describing the center's function. The center offers licensing of available EPA patents. ETC2 also helps environmental companies find funding from venture capitalists and private investors as well as helping small businesses obtain funding from other government programs.

MAIN RESOURCES OFFERED R1, R3, R4, R5, R6, R8, R12

TECHNOLOGY FUNCTIONS Various

SECTOR RELEVANCE Various

CONTACTS

For licensing of technology or performing research through a CRADA, e-mail: epaetc2@battelle.org

Submit questions at www.etc2.org/contact.html

ETC2

20445 Emerald Parkway Drive

Cleveland, OH 44135

Tel: 216-898-6400

Fax: 216-898-6500

E-Mail: epaect2@battelle.org

Small Business Gateway

WEB SITE www.epa.gov/smallbusiness/index.htm

KEY WORDS drinking water, ecosystem, endocrine disruptors, particulates

PURPOSE to provide a one-stop portal for environmental resources offered by the EPA and other government agencies to small businesses.

SUMMARY *The Gateway* offers support to small businesses, including grants to small high-tech firms for research and development of cutting-edge technology. A large portion of the Gateway is devoted to R&D and commercialization of new technology through the Small Business Innovative Research (SBIR) program (See Chapter 4).

MAIN RESOURCES OFFERED R2, R3, R6, R12, R14

TECHNOLOGY FUNCTIONS Various

SECTOR RELEVANCE small business, research and development

CONTACTS

SBIR proposal information:
E-mail: Johnson.marsha@epa.gov

Information on loans, grants, financial and paperwork assistance: www.sba.gov/regions/states.html

Office of the Chief Financial Officer

Environmental Finance Program

WEB SITE www.epa.gov/efinpage/

KEY WORDS financing, pollution prevention, technical assistance

PURPOSE to assist communities in funding environmental projects and to lower costs and increase investments associated with the environment.

SUMMARY *EFP* provides financial technical assistance to the regulated community. The program creates partnerships with state and local governments and the private sector to fund creative approaches to environmental needs and build capacity to manage the cost of environmental services. *A Guidebook of Financial Tools* lists hundreds of financing tools that federal, state and local governments and the private sector can use to meet environmental mandates and create sustainable systems. The program's university-based environmental finance centers (EFCs) educate state and local officials and businesses on lowering costs of compliance, pollution prevention and other financial issues. The Environmental Financing Information Network (EFIN) provides information on financing alternatives for state and local environmental programs and small businesses, including referrals to experts. EFP includes an information hotline, abstracts of case studies and the *EFAB Newsletter.*

MAIN RESOURCES OFFERED R1, R2, R4, R10, R11, R12, R14

TECHNOLOGY FUNCTIONS Various

SECTOR RELEVANCE environmental services, small business, state and local government

CONTACTS

US EPA
Environmental Financing Information Network, (Mail Code) 2731R
Ariel Rios Building
1200 Pennsylvania Avenue, NW
Washington, DC 20460
Infoline: 202-564-4994;
Fax: 202-565-2587
E-mail: efin@epa.gov

For the Guidebook of Financial Tools: www.epa.gov/efinpage/guidebooktp/index.htm
E-mail: mcprouty.timothy@epa.gov

Office of Policy, Economics and Innovation (OPEI)

WEB SITE www.epa.gov/opei

KEY WORDS cleanup, conservation, industrial waste, pollution prevention, recycle/reuse

PURPOSE to support EPA's mission through economic analysis and by promoting innovation to achieve better, more cost-effective environmental and public health protection.

PROGRAMS

- Sustainable Industry Development Program
- Community-Based Environmental Protection

SUMMARY OPEI supports the integration between regulatory approaches and voluntary actions that encourage environmental responsibility. Its *Partners for the Environment* program helps businesses and other organizations to cut costs, prevent pollution and conserve natural resources. Located on its publications Web site are numerous publications of reports, addresses and symposia from the last 3 years that OPEI, its programs and other EPA programs have worked toward environmental protection. OPEI also runs a listserve, which sends e-mails to its members every 2 weeks. Each e-mail, as well as providing a link to obtain more information, includes new reports, grant opportunities, technologies and innovations that the EPA supports.

The *Sustainable Industry Development Program* works with selected industries to provide incentives and remove barriers to companies so that they can devise new and better ways to reduce pollution and conserve resources. The *Community-Based Environmental Protection* program (CBEP) tries to protect the environment by focusing on the health of an ecosystem and the humans that live within the boundaries of that ecosystem. CBEP is concerned with specific places that contain environmental hazards rather than the individual hazard itself.

MAIN RESOURCES OFFERED R1, R2, R5, R10

TECHNOLOGY FUNCTIONS T1, T4, T8, T13

SECTOR RELEVANCE Various

CONTACT

Office of Policy, Economics and Innovation (1804A)
U.S. Environmental Protection Agency
Ariel Rios Building
1200 Pennsylvania Avenue NW
Washington, DC 20460
Tel: 202-564-4332
E-mail: opei@epa.gov

To subscribe to listserve, email: listserver@unixmail.rtpnc.epa.gov
In the message type: subscribe innovation YOUR NAME

Sustainable Industry Development Program

WEB SITE www.epa.gov/sustainableindustry/

KEY WORDS air quality, chemicals, commercial waste, environmental assessment, industrial waste, metals, pollution prevention, recycle/reuse

PURPOSE to partner with industry sectors to identify incentives that could be used and barriers that could be removed to improve environmental management capabilities

SUMMARY The *Sustainable Industry Development Program* is a new approach to EPA policy development involving cooperation with industry's decision makers. The goal is to gain a better understanding of the reasons that businesses act either against or with the government in its efforts to protect the environment. By identifying sectors of industry that have opportunities for further progress in reducing or preventing pollution, the EPA and industry join forces to devise solutions. Then, pilot projects are conducted and stakeholders evaluate the results (see metal finishing example). Industry sectors working within the program are listed under the heading Sector Relevance.

MAIN RESOURCES OFFERED R1, R5, R9, R10, R12, R13, R14, R15

TECHNOLOGY FUNCTIONS T1, T2, T4, T6, T8, T13

SECTOR RELEVANCE chemical, food processing, metal casting, metal finishing, photo processing, ship building and ship repair, travel and tourism
Sample Project

Metal Finishing 2000 is one of EPA's sustainable industry projects in which industry environmental performance leaders meet defined program criteria and pursue industry-wide performance goals developed by the metal finishing sector in order to receive operational flexibility. One of the necessary criteria is 1 to 3 years of demonstrated good compliance with federal, state and local environmental regulations. An example of a goal that must be achieved to receive the benefits of the program is a 50% reduction in metals emissions to water. Benefits received include reduced reporting on compliance, quicker facility permits for improvements and a grace period to test new equipment.

CONTACTS

Bob Benson
Director, Industry Sector Policy Division
Tel: 202-260-8668
E-Mail: Benson.robert@epa.gov

Industry Sector Policy Division
Tel: 202-260-8661,
Fax: 202-260-8662

Community-Based Environmental Protection (CBEP)

WEB SITE www.epa.gov/ecocommunity

FIGURE 6.3
Community-based projects involve citizens like these volunteers cleaning up a waterway in Louisiana's Lake Pontchartrain Basin (see project description below).

KEY WORDS climate, conservation, ecological assessment, ecosystem, geographic information system, habitat, lakes and rivers, nonpoint pollution abatement, site assessment, sustainability, watershed

PURPOSE to work collaboratively with other federal, state, tribal and local governments as well as with voluntary citizen action to address the environmental issues not subject to traditional EPA regulatory approaches.

SUMMARY *Community-Based Environmental Protection* differs from traditional EPA programs by seeking to improve specific ecosystems that have health hazards, instead of focusing on the hazard itself. The CBEP approach helps communities identify problems, set goals and achieve results using the most appropriate solutions for that specific community. This community-based approach fosters partnerships on all levels of government, as well as with industry, public interest groups, private landowners and others. Some of the key principles of CBEP are: focusing on a defined geographic area, working collaboratively with a wide range of stakeholders through partnership and assessing, protecting and restoring the quality of air, water, land and living resources. Typical CBEP stakeholders include EPA and state officials, nongovernmental organizations, city planners and community groups. In many places, EPA supports and assists the efforts of others by providing environmental information, monitoring systems, scientific analysis and other types of assistance. EPA has offered grant programs in the past, but, as of the fiscal year 2001, the grants have been suspended. The Web site offers a range of publications about environmental issues in communities including a CD-ROM entitled *This Place Called Home: Tools for Sustainable Communities*. The Web site also has a link to tools such as reports on the economics of making a community more environmentally sound, and links to other Web sites on sustainable development. The site also links to a number of case studies from the last decade.

Sample Project

Lake Pontchartrain Basin, Louisiana, and its adjoining lakes make up one of the largest estuaries in the United States. About 1.5 million people live in the 4700 sq. mi. watershed basin in southeastern Louisiana. The Lake Pontchartrain Basin Foundation and the U.S. congress initiated the project. The basin had many environmental problems, including sewage from fishing camps, poorly sewered and non-sewered communities, loss of wetlands, diminished shellfish and fish harvests and habitat destruction from urban development. Remediation included such elements as stormwater treatment (using created wetlands and retention ponds) and restoration of submerged aquatic vegetation.

MAIN RESOURCES OFFERED R1, R5, R10, R14, R15

TECHNOLOGY FUNCTIONS T1, T2, T3, T10, T11, T12, T13

SECTOR RELEVANCE Varies according to community

CONTACTS Regional contact information on Web site.

Karen Young
U.S. EPA Region VI
1445 Ross Avenue
Dallas, TX 75202-2733
Tel: 214-665-6689
Fax: 214-665-6679

Carlton Dufrechou
Lake Pontchartrain Basin Foundation
P.O. Box 6965
Metairie, LA 70009
Tel: 504-836-2215
Fax: 504-836-7283

Office of Pollution Prevention and Toxics (OPPT)

WEB SITE www.epa.gov/opptintr

KEY WORDS see individual programs

PURPOSE to promote pollution prevention as the guiding principle for controlling industrial pollution at its source; safer chemicals, risk reduction to minimize exposure to substances such as lead, asbestos, dioxin and polychlorinated biphenyls and public understanding of risks.

PROGRAMS

- Enviroene
- Design for the Environment (DfE)
- Green Chemistry
- Green Engineering

SUMMARY OPPT has responsibility for administering the following laws:

- Asbestos Hazardous Emergency Response Act (AHERA)
- Asbestos School Hazard Abatement Act (ASHAA)
- Emergency Planning and Community Right-to-Know Act (EPCRA)
- Pollution Prevention Act (PPA)
- Toxic Substances Control Act (TSCA)

The office is the main source of pollution-prevention information for industry and for communities with exposure to chemical and toxic substances. Numerous tools (reports, case studies, software) are available for use by industry and state and local government regulators.

EPA has, for example, developed a Web-based P2 Framework, software for risk screening that incorporates pollution-prevention principles in the design and development of chemicals. *Green Products* promotes environmentally preferable purchasing and safe labeling to encourage the consideration of environmental factors in the purchasing and use of products. *Business Practices Initiatives* include environmental assistance to small businesses, P2 Finance, the Voluntary Standards Network and the Environmental Accounting Project. *Design for Environment (DfE)* and *Green Chemistry* are industry partnerships that foster design and process modifications to reduce negative environmental impacts in specific industries (see Sector Relevance). DfE also houses the *Enviroene* program, found separately in this book. *P2 Grants* promote incentives for adoption of P2 concepts and practices at the state, tribal and local level. The *Pollution Prevention Information Clearinghouse* has detailed information and documentation on all P2 programs.

MAIN RESOURCES OFFERED R1, R2, R7, R10, R11, R12, R13, R14, R15

TECHNOLOGY FUNCTIONS Various

SECTOR RELEVANCE public interest groups, state and local government, universities, various industries

CONTACT

Jim Bradshaw (OPPT/IMD/IAB)

Tel: 202-260-1543
E-mail: www.bradshaw.jim@epamail.epa.gov

Enviroene (OPPT)

WEB SITE es.epa.gov/

KEY WORDS compliance, enforcement, pollution prevention, remediation

PURPOSE to promote a single source for pollution prevention, compliance assurance, and enforcement information and databases.

SUMMARY *Enviroene*, the EPA's primary pollution-prevention Web site, links to a variety of environmental sources inside and outside the agency. For example: Partners for the Environment, composed of thousands of organizations working with the EPA to set and reach conservation and pollution prevention goals; the National Pollution Prevention Roundtable, which sponsors workgroups and conferences; EPA Sector Notebooks, a series of profiles containing environmental information on many major industries; and the Joint Service P2 Technical Library, a Defense Department source of information on pollution prevention and environmental compliance problems, as well as links and phone numbers to contacts. Updated daily, Enviroene offers three Internet information groups (federal, state or local and international) that provide access to online information on pollution prevention (general and sector-based), cleaner production resources and compliance assistance resources. The site also has case studies and links to VendInfo, which contains more than 400 listings of pollution-prevention equipment, products and services. Vendors can submit their technology to the clearinghouse and the search engine distinguishes between verified and unverified technologies. Additional offerings include information hotlines, bulletin boards, clearinghouses, training resources, environmental laws and newsletters.

MAIN RESOURCES OFFERED R1, R2, R4, R5, R7, R14

TECHNOLOGY FUNCTIONS T1, T2, T3, T4, T5, T6, T7, T8, T9, T10, T11, T12

SECTOR RELEVANCE Various

CONTACTS

Multiple directories and yellow pages:
es.epa.gov/new/contacts/contacts/contacts.html

Sample Technology: Low Energy Recycling of Foundry Sand

WEB SITE www.es.epa.gov/techinfo/case/michigan/mich-cs4.html

KEY WORDS foundry sand

ENVIRONMENTAL CHALLENGE Sand is used in metal foundries for mold and core production. Like other metal foundries, Wolverine Bronze Company has seen the costs of buying new sand and disposing of used sand rise steadily over the past several years. Traditional sand-reclamation systems use high levels of energy to heat sand during cleaning, as well as when the system is idle. These recycling systems are almost as expensive as new sand and waste a great deal of energy.

SECTOR RELEVANCE metal casting

DEVELOPER Tanoak Enterprises, Inc., of Barrie, Ontario, Canada, designed this system for Wolverine Bronze foundry.

TECHNOLOGY DESCRIPTION At the heart of the system installed at Wolverine Bronze is an attrition unit in which sand grains rub together at high speeds. The sand attrition unit causes two sand streams traveling at different speeds and flowing in opposite directions to repeatedly collide. Residual binder and inorganic contaminants are scrubbed from the sand grain surfaces by the abrasion. The resulting cleaned sand is almost like virgin sand. Motion, temperature, electrical power, pressure and limit sensors provide automated monitoring of the system. It complies with federal and state environmental requirements. This process provides a sand product that is suitable for both mold and core production.

PROJECT RESULTS The system produces 4 tons of recycled sand per hour in 700 lb. batches. The recycled sand costs $9.72 per ton to reclaim, compared with the traditional treatment price of $30 per ton. Typically, the system operates 13 hours per day, usually at night and on weekends when power rates are lowest.

CONTACT

Information on low energy sand reclamation systems for foundries:
Office of Waste Reduction Services
Environmental Services Division
Departments of Commerce and Natural Resources
P.O. Box 30004
Lansing, Michigan 48909
Tel: 517-335-1178
1-800-NO-2-WASTE (Toll Free in Michigan)

Design for the Environment (DfE)

FIGURE 6.4
EPA's Design for the Environment. (DfE) program involves industry sectors in efforts to promote cleaner, cheaper and smarter business practices.

WEB SITE www.epa.gov/opptintr/dfe/

KEY WORDS design for the environment, energy conservation

PURPOSE to integrate health and environmental considerations in business decisions.

SUMMARY To address challenges of economic and environmental sustainability, the EPA's *DfE Program* is building on the concept pioneered by industry. The primary focus of DfE is to encourage businesses to incorporate a complete life-cycle perspective of environmental considerations into the redesign of products, processes and management practices. DfE supports the shift from a reliance on end-of-pipe pollution control to prevention and encourages the development and use of innovative technologies. DfE partnerships provide reliable health and environmental information to businesses that allows them to make environmentally preferable business decisions. Benefits of sustainable business practices include cost savings, reduced business and environmental risks, as well as expanded business and market opportunities. The DfE program translates informational materials into other languages to reach target audiences, including Spanish and Korean. The *DfE Integrated Environmental Management System (IEMS)* guide and template is actively in use in countries around the globe.

Some of the program's key accomplishments include:

- In the drycleaning industry, promoting the development of innovative new technologies, such as the use of water and liquid CO_2 as alternative cleaning solvents that have reduced the use of perchloroethylene by more than 68%.
- In the electronics industry, identifying alternative processes and chemical usage for conductivity in circuit boards, emerging technologies in computer display screens and lead-free surface coating alternatives to the standard hot air solder leveling process. DfE electronics partnerships have reduced formaldehyde use by 240,000 lbs and saved over 400 million gals of water and 15 billion Btus of energy consumption per year.
- In the industrial-laundry industry, developing chemical substitutes that have led to significant reductions in toxic chemical, water and energy usage.

MAIN RESOURCES OFFERED R1, R2, R10, R14, R15

TECHNOLOGY FUNCTIONS T1, T2, T4, T6, T8, T11, T12

SECTOR RELEVANCE Industries: adhesives, auto refinishing, cleaning-product formulators, computer displays, flexography, garment and textile care, laundries, lithography, metal finishing, printing, printed wiring boards, screen printing. Public sector: nonprofits, research institutions, universities.

CONTACTS

U.S. Environmental Protection Agency
Design for the Environment Program
Tel: 202-564-8780
E-mail: oppt.dfe@epamail.epa.gov
Environmental Accounting Project
www.epa.gov/opptintr/acctg/

Green Chemistry

WEB SITE www.epa.gov/greenchemistry/

KEY WORDS chemical catalysis, chemical synthesis, feedstocks, solvents, synthetic reactions, toxic substances,

PURPOSE to promote innovative chemical technologies that reduce or eliminate the use or generation of hazardous substances during the design, manufacture and use of chemical products and processes.

SUMMARY As defined by EPA, green chemistry encompasses all aspects and types of chemical processes that reduce negative impacts to human health and the environment relative to the current state of the art. The *Green Chemistry* program fosters the research, development and implementation to develop products and processes that are more environmentally benign. It operates through voluntary non-regulatory partnerships between the government and academia, industry, scientific societies, trade organizations, national laboratories, research centers and other government agencies. The annual Technology for a Sustainable Environment awards (sponsored with the National Science Foundation) typically range from $50,000 to $150,000. Academic and nonprofit institutions, along with state and local governments, are eligible. The Green Chemistry Expert System (GCES) allows users to build a green chemical process, design a green chemical, or survey the field of green chemistry. Both new and existing chemicals and their synthetic processes are included. The Web site also has a calendar of upcoming and past green chemistry conferences.

MAIN RESOURCES OFFERED R2, R4, R6, R10, R12, R14

TECHNOLOGY FUNCTIONS T1, T2, T6

SECTOR RELEVANCE academia, nonprofit, state and local government, trade associations

CONTACTS Technology for a Sustainable Environment solicitation:

Steve Lingle at 202-564-6821 or Barbara Karn at 202-564-6824
(See also National Science Foundation, NSF)
Research Grant Opportunities Fact Sheet
www.epa.gov/greenchemistry/docs/grants.pdf
EPA's Pollution Prevention Clearinghouse
Tel: 202-260-1023.

Green Engineering

WEB SITE www.epa.gov/opptintr/greenengineering/index.html

KEY WORDS chemical engineering, feedstocks, industrial ecology, risk assessment, sustainable technology, waste streams

PURPOSE to incorporate risk-related concepts into chemical processes and products designed by academia and industry.

SUMMARY The Web site offers case studies of technologies designed to produce less pollution and a risk-based suite of tools for assessing chemical hazards in process design. Extensive outreach programs are conducted with organizations such as the American Institute of Chemical Engineers (AIChE) and the American Society for Engineering Education (ASEE). The Chemical Engineering Summer School for professors organized by ASEE is held every 5 years (see ASEE for the 2002 workshop details). The Web site welcomes ideas, feedback and comments as well as submissions of new technologies and case studies. It also has links to workshop partners and pollution prevention Web sites.

MAIN RESOURCES OFFERED R1, R4, R10, R11, R12, R14

TECHNOLOGY FUNCTIONS T1, T2, T6

SECTOR RELEVANCE chemical manufacturers, engineering schools

CONTACT

Sharon Austin
Green Engineering Program Coordinator
U.S. Environmental Protection Agency; Chemical Engineering Branch (7406)
1200 Pennsylvania Avenue, NW
Washington, DC 20460
Tel: 202-260-2367
E-mail: green-engineering@epa.gov

Sample Technology: Mi 'nizing Hazardous Chemicals in the Paper Industry: the POM process

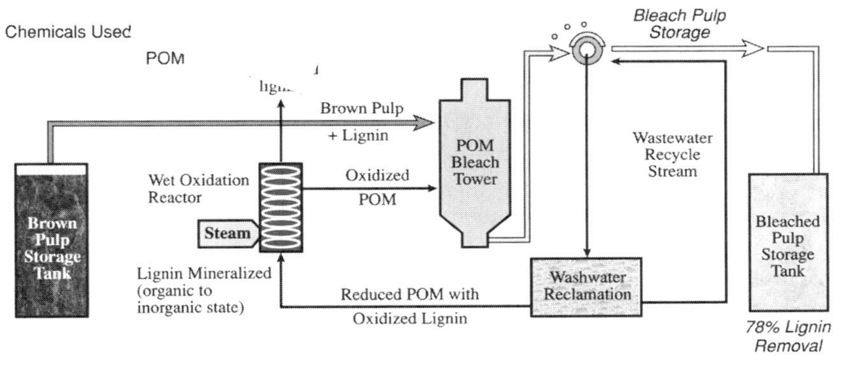

FIGURE 6.5
Green delignification process for paper pulp processing substitutes polyoxometalates (POM) for chlorine bleach. Use of the latter requires treatment of toxic effluents. In the new process only evaporator condensate, mainly ethanol, ererges (Steve Hall, after EPA, with permission).

KEY WORDS bleaching agents, chlorine, delignification, dioxins

ENVIRONMENTAL CHALLENGE EPA regulations call for the reduction or elimination of chlorine and byproduct discharges. Chlorine is commonly used in paper processing and bleaching to dissolve lignin, the material that holds cellulose fibers together in wood pulp. Once the lignin is washed away, the pulp can be turned into paper. Chlorine is then used again to make the pulp white. When wood pulp is bleached, the chlorine, lignin and cellulose fibers react to produce highly toxic substances, including dioxins, which accumulate in the air, water and soil. Dioxins have been linked to many ailments, including cancer, deformities and developmental problems in children, and reproductive disorders.

PROJECT RESULTS While the delignification process completely eliminates bleach plant effluent, it requires increased steam and energy usage as well as new capital equipment. The POM process also requires a higher process flow temperature. The increased capital cost in mills already built prohibits immediate implementation. Engineers will soon be moving the POM process to the pilot plant phase, while continuing to research and test the final product. The challenge will be to reduce process temperature and to eliminate the oxidative reactor to decrease steam requirements.

TECHNOLOGY DESCRIPTION Carl Houtman of the USDA Forest Product Service and of the University of Wisconsin at Madison conducted a case study on minimizing the use of hazardous chemicals in the paper industry. To eliminate the use of and exposure to hazardous chemicals in the bleaching process, a new delignification agent (a polyoxometalate or POM) is used to provide the basis for closed-mill bleaching technology. This technology eliminates the use of hazardous chemicals while maintaining effective lignin removal.

SECTOR RELEVANCE pulp and paper mills

CONTACT

Green Engineering Comment Page: www.epa.gov/opptintr/greenengineering/comments.html

E-mail: green-engineering@epa.gov

Office of Research and Development (ORD)

WEB SITE www.epa.gov/ord/

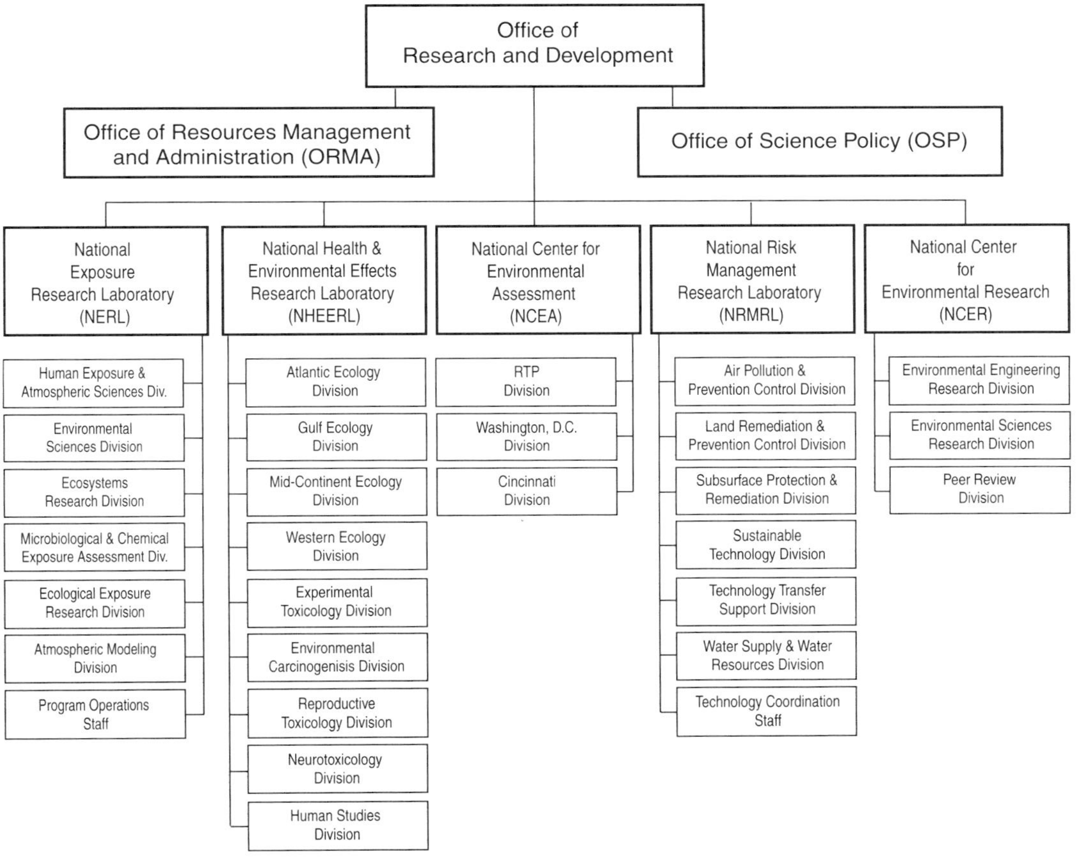

FIGURE 6.6
Results from ORD's labs and centers help communities and organizations incorporate research and technology into environmental management activities (ORD/EPA, with permission).

KEY WORDS see NERL and NRMRL entries

PURPOSE to conduct leading-edge research and foster the sound use of science and technology to protect human health and safeguard the natural environment.

SUMMARY *ORD* is the principal scientific and research arm of EPA, but also provides extensive technology-transfer information for all environmental media through various projects and programs. The research is carried out by in-house and extramural research from three national laboratories and two national centers located in a dozen facilities around the country and in Washington, D.C. (for locations and descriptions of each function, see map and organization chart above). An example of ORD's technology transfer resources is the *Watershed Atlas (see* Contacts). The Atlas provides geospatial displays, analyses of information and data on watersheds across the nation. Users can search by such elements as watershed name, themes (e.g., downstream effects, stressors and where restoration and protection projects are underway) and condition reports.

MAIN RESOURCES OFFERED R1, R2, R3, R5, R6, R11, R12, R13, R14, R15

TECHNOLOGY FUNCTIONS Various

SECTOR RELEVANCE Various

CONTACTS

Content Specialist
E-mail: David.Richard@epa.gov
Main Telephone Number:
Tel: 202-564-6620
Watershed Atlas
www.epa.gov/wateratlas/

Environmental Monitoring for Public Access and Community Tracking (EMPACT)

WEB SITE www.epa.gov/empact

KEY WORDS environmental data

PURPOSE to provide timely environmental information to communities across the nation.

SUMMARY Housed in EPA's Office of Environmental Information, *EMPACT* is designed to work with the largest metropolitan communities to collect, manage and present environmental information to the public so that communities and individuals can make informed day-to-day decisions. EPA and the communities use a variety of dissemination methods such as the Internet, city kiosks, television, radio, newspapers, fliers, billboards, town hall meetings, community organizations and person-to-person communication. Two rounds of grants for projects in more than 90 communities have been funded, but no funds are expected to be available for FY 2002. Types of environmental data to be collected and provided, sometimes on a daily basis, include air and water, ultraviolet radiation exposure, the risk of lead exposure to children and contamination at hazardous-waste sites, as well as various technology transfer products for community monitoring (see Contacts).

Sample Project

Jefferson Parish Project (Louisiana): Along the coastline of the Gulf of Mexico, wetlands are being lost at an alarming rate and hypoxia in the Gulf is a growing local and national concern. To restore coastal wetlands, fresh water and sediments from the Mississippi River must be diverted, but there are concerns about the impact that nutrient-rich river water may have on water quality and growths (blooms) of phytoplankton. The Jefferson Parish EMPACT project utilizes satellites to monitor freshwater diversions and harmful algal blooms (HABs) in Lake Salvador and other waters in the New Orleans Area. A data-collection station in Lake Salvador will transmit real-time water quality and weather information to the U.S. Geologic Survey Louisiana Hydrowatch Web site. Water samples taken from Lakes Salvador and Cataouche will be analyzed to verify interpretation of the satellite images. Other water quality monitoring programs in the New Orleans metropolitan area will be integrated into the project as appropriate.

MAIN RESOURCES OFFERED R1, R2

TECHNOLOGY FUNCTIONS T1, T2, T3, T5, T6, T7, T9, T11

SECTOR RELEVANCE local government

CONTACTS

Dr. Denise Shaw
EMPACT Program Director
Tel: 202-564-3234
Publications
Tel: 1-800-490-9198 or 513-489-8190
Fax: 513-569-7566)Tel:

Environmental Technology Verification Program (ETV)

WEB SITE www.epa.gov/etv/

KEY WORDS Various

PURPOSE to verify the performance of innovative technical solutions (and) substantially accelerate the entrance of new environmental technologies into the domestic and international marketplace

SUMMARY The *Environmental Technology Verification Program* is an independent, objective source of performance information. Through government and private testing partnerships, the ETV program evaluates the performance of commercially available environmental technology in all media, with pilots developed to test technology for air, water, monitoring and pollution prevention. Any technology vendor is invited to participate. Market input is collected from groups of stakeholders, consisting of technology buyers, sellers, permitters and others. Test plans and protocols are developed by technical experts, stakeholders and vendors and are then peer reviewed. ETV has outreach programs across the country and overseas to not only recruit technology vendors and verify their creations, but also to help state and local permitters feel more comfortable with and accepting of ETV's results. Lists of technologies verified (searchable by media) or undergoing pilot testing are available on the ETV Web site, along with performance reports. The "What's New" link leads to information on upcoming workshops, the latest verification test results, a glossary of ETV terms and postings of technologies wanted for current testing. The site also has links to ETV partners, such as the Civil Engineering Research Foundation (see next entry) that collaborate in verifications.

MAIN RESOURCES OFFERED R1, R4, R9, R10, R12, R14

TECHNOLOGY FUNCTIONS Various

SECTOR RELEVANCE Various

CONTACTS

ETV technology staff: bauer.sarah@epa.gov
conley.tina@epa.gov

Environmental Technology Evaluation Center (EvTEC)

Civil Engineering Research Foundation (CERF)

WEB SITE www.cerf.org/evtec/index.htm

KEY WORDS bioremediation, concrete repair, erosion control, storm water management, transportation systems, wastewater sludge

PURPOSE to produce independent, market-based technology verification and thus accelerate the adoption of environmental technologies.

FIGURE 6.7
Jensen Precast Stormvault™ A water quality capture volume treatment system catches oil, grease and fine sediment from stormwater runoff. It can help restore pre-development runoff rates by significantly reducing erosion and stream degradation due to increased flows from urbanization (Jensen Precast, with permission).

SUMMARY CERF and the EPA cooperated to initiate *EvTEC*, an independent body working with the EPA's Environmental Technology Verification program (ETV). EvTEC verifies the performance of new, commercially ready environmental technologies and transfers this information to those who recommend, specify, approve, or purchase these technologies, such as federal and state regulators, facility owners and operators and consulting engineers. EvTEC's market-based evaluations are conducted by the private sector, including those who regulate, authorize, approve, purchase, or use the technology in question. Technical reports describing EvTEC evaluations can be ordered online for a fee, while brief descriptions of some evaluations are posted at no charge. Ongoing tests and future evaluations are listed, with announcements of products and venders that can participate. A monthly electronic newsletter, *EvTEC Edge*, is also available.

MAIN RESOURCES OFFERED R1, R4, R9, R10, R12, R13, R15 (economic assessment)

TECHNOLOGY FUNCTIONS Various

SECTOR RELEVANCE contractors, engineering consultants, federal regulators, food processing, metal finishing, pulp and paper, state regulators, textiles

CONTACT

EvTEC: www.cerf.org/evtec/contact.htm
Order Technical Reports: 800-548-2723 or 703-295-6300
E-mail: marketing@asce.org

Sample Technology: Silt Fence Machine (EvTEC)

WEB SITE www.cerf.org/evtec/EVAL/TOMMY.HTM

KEY WORDS nonpoint source, runoff, silt fencing

ENVIRONMENTAL CHALLENGE Traditional fencing methods do not always compact soil surrounding the fence, thus allowing runoff water to saturate this soil. This leads to washouts under the fence, a major traditional silt fence installation problem.

FIGURE 6.8
The Tommy Silt Fence Machine aids soil retention at construction sites (Carpenter Erosion Control, with permission).

SECTOR RELEVANCE construction.

TECHNOLOGY DESCRIPTION The Tommy Silt Fence Machine slices through soil while inserting silt fencing, with no displacement of soil. Silt fencing is pulled off a roll by the wheel, funneled into the machine and inserted into the soil being held open by the panels. As the machine progresses, soil collapses onto the fabric, securing the fence in the desired position. By disrupting the soil during fence installation only slightly, the Tommy creates optimal soil conditions for compaction without losing moisture. This prevents runoff water from saturating the soil next to the fabric, thus eliminating washouts under the fence.

PROJECT RESULTS Testing for the Tommy was completed in August 2000. The machine is currently in use in more than 25 states. It works in all types of soil, from fractured rock and compacted clays to wetlands and on slopes. The Tommy also has lower maintenance and repair costs than traditional methods due to its simple design. Verification report available at www.cerf.org/evtec/news/reports.htm

CONTACTS

Carpenter Erosion Control
3718 SW Court Avenue
Ankeny, IA 50021
Tel: 515-965-4665
Fax: 515-964-0151
E-mail: tom@tommy-sfm.com
EvTEC
Brian Rustia
Tel: 202-785-6432
E-mail: brustia@cerf.org

National Exposure Research Laboratory (NERL)

WEB SITE www.epa.gov/nerl/

KEY WORDS agriculture, alternative migration, climate change, exposure assessment, exposure management, land use, receptor analysis, stressor analysis, watershed management

PURPOSE to reduce and quantify the uncertainty in EPA's exposure and risk assessments for all environmental stressors

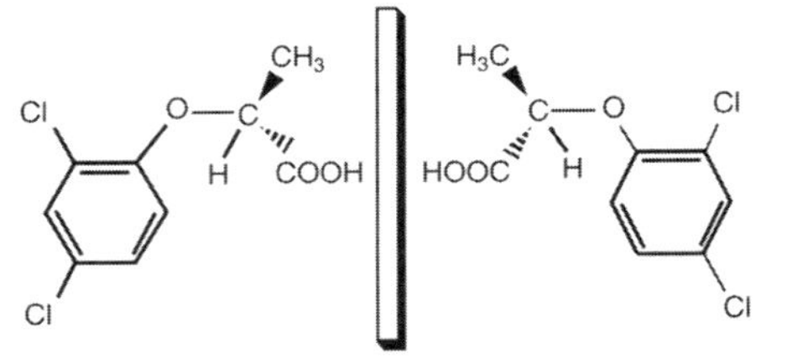

FIGURE 6.9
Up to 25% of pesticides (as well as other pollutants like PCBs) are chiral molecules (mirror images of each other) that can connect in two different ways—one toxic, one not. Research is exploring the prospect of designing herbicides and pesticides with less environmental impact.

SUMMARY The *NERL* conducts in-house and extramural research to develop measurements and models to assess and predict the exposure of humans and ecosystems to pollutants in air, water, soil and food. It provides educational materials and technical support to EPA programs, state and local or tribal governments and other federal agencies to help in performing state-of-the-science exposure assessments. Research abstracts are written for decision makers, nonscientific personnel and the general public, backed up by technical reports. The lab's Center for Exposure Assessment Modeling distributes and supports a variety of multimedia and ecosystem response models and the Web site links to a model database.

MAIN RESOURCE OFFERED R1, R2, R4, R11, R12, R13, R14, R15

TECHNOLOGY FUNCTIONS T1, T2, T3

SECTOR RELEVANCE agriculture, chemicals, nonprofit organizations, small business, state and local government, students, universities, utilities

CONTACTS

Robin Bailey, NERL Public Outreach Specialist
Tel: 919-541-7906
E-mail: baily.robin@epa.gov

Publications:
www.epa.gov/ORD/publications/
Tel: 919-541-4871
E-mail: obriant.mary@epamail.epa.gov

National Risk Management Research Laboratory (NRMRL)

WEB SITE www.epa.gov/ORD/NRMRL/

KEY WORDS dechlorination, fuel cells, lead, ozone, resource recovery, volatile organic compounds, waste

PURPOSE conduct research preventing and reducing risks from pollution threatening human health and environment.

SUMMARY *NRMRL* collaborates with both public and private partners to develop technologies that reduce the cost of pollution compliance and anticipate environmental problems. It also provides technical assistance to communities in environmental compliance and strategies. The Sustainable Technology Division develops and demonstrates pollution prevention tools. The Clean Processes Branch lists technologies for pollution prevention as well as resource recovery and reuse. The Industrial Multimedia Branch develops and evaluates approaches to reduce industrial pollution and the Systems Analysis Branch focuses on cost-effective decision-making tools that integrate environmental solutions, life-cycle concepts and economics. The lab also manages EPA's Superfund Innovative Technology Evaluation (SITE) program (see the following listing) and Environmental Technology Verification (ETV) program (described elsewhere in this chapter).

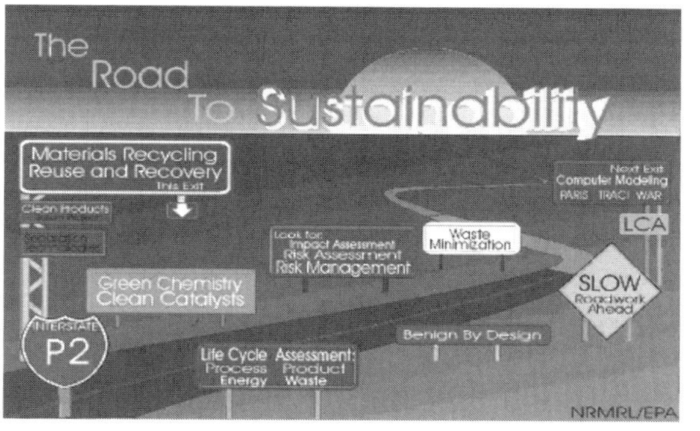

FIGURE 6.10
EPA's Risk Management Lab addresses many pollution prevei opportunities (NRML, with permission).

MAIN RESOURCES OFFERED R1, R4, R10, R11, R12, R13

TECHNOLOGY FUNCTIONS T1, T2, T3, T4, T5, T6, T7, T8, T12

SECTOR RELEVANCE various, including chemicals, metal forming and finishing, mining

CONTACTS

Clean Processes Branch
Teresa Harten
Tel: 513-569-7565

Industrial Multimedia Branch
Roger Wilmoth
Tel: 513 569-7509
Fax: 513-569-7471
E-mail: wilmoth.roger@epa.gov

Sustainable Environments Branch
Heriberto Cabezas
Tel: 513-569-7350

Systems Analysis Branch
Gordon Evans
Tel: 513-569-7684
Fax: 513-569-7111
E-mail: evans.gordon@epa.gov

Superfund Innovative Technology Evaluation (SITE)

WEB SITE www.epa.gov/ORD/SITE/

KEY WORDS air quality, bedrock, groundwater, hazardous waste, *in situ*, lead, soil, surface water, sediment

PURPOSE to promote the development, commercialization and implementation of cost- and performance-effective hazardous waste remediation technologies.

SUMMARY Since the initiation of the *SITE* program in 1986, cleanup of contaminated sites through the use of innovative technologies has resulted in a total inflated cost savings of more than $2.1 billion. Recently, the program has shifted its emphasis from technologies requiring the removal of soil or groundwater (*ex situ*) to *in situ* technologies that treat contamination in place. The program does not give financial support for remediation or technology development, but sponsors SITE demonstrations to assess the performance of the technology, potential need for pre- and post-processing of the waste, applicability to other sites, potential operating problems and approximate capital and operating costs. Innovative technology evaluation reports are available online. Other resources include quarterly project reports, a superfund videotape library and a CD-ROM database of the SITE technology profiles searchable by parameters such as media, contaminants and technology vendors.

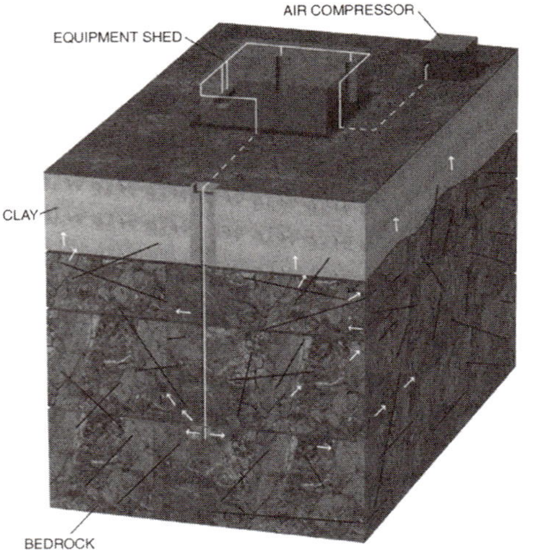

FIGURE 6.11
Bioremediation in bedrock involves stimulating growth of in-situ bacteria (courtesy of ITT Night Vision/Earth Tech).

MAIN RESOURCES OFFERED R1, R2, R4, R6, R10, R11, R12, R13

TECHNOLOGY FUNCTIONS T1, T7, T9, T10, T11

SECTOR RELEVANCE various, including chemicals, environmental remediation, environmental consulting, mineral processing

CONTACT

Bill Frietsch
E-mail: frietsch.william@epa.gov
SITE demonstrations:
Annette Gatchett
Tel: 513-569-7697

For information on ORD's Grants and Fellowships: www.es.epa.gov/ncerqa

For CD-ROM: Tel: 800-490-9198 (EPA/540-C-99/500)

Office of Solid Waste and Emergency Response (OSWER)

WEB SITE www.epa.gov/swerrims/

KEY WORDS brownfields, chemical emergency, emergency response, oil spills, Superfund, underground storage tank

PURPOSE to provide policy, guidance and direction for the Agency's solid waste and emergency response programs.

PROGRAMS

- Chemical Emergency Preparedness and Prevention (CEPPO)
- Technology Innovation Office (TIO)
- Underground Storage Tanks (OUST)

SUMMARY *OSWER* furnishes technical assistance in the development, management and operation of solid-waste activities and analyzes the recovery of useful energy from solid waste. Another program covers abandoned and active hazardous-waste sites and accidental release (including some oil spills) as well as the encouragement of innovative technologies for the remediation of contaminated soil and groundwater and Brownfields development. Most of these programs are compliance-focused but contain some information on best practices and technology innovations for source reduction, recycling, combustion and land disposal. Grants, networking and information are available in the *Jobs Through Recycling (JTR)* program, which targets businesses that want to process or manufacture recycled content products.

MAIN RESOURCES OFFERED R1, R2, R4, R5, R6, R12, R13, R15

TECHNOLOGY FUNCTIONS T1, T2, T3, T4, T5, T6, T7, T8, T9, T10, T11, T12, T13

SECTOR RELEVANCE Various

CONTACT

See links for individual programs.

Chemical Emergency Preparedness and Prevention Office (CEPPO)

WEB SITE www.epa.gov/swercepp/index.html

KEY WORDS chemical emergency, emergency response, risk management

PURPOSE to provide leadership, advocacy and assistance to: (1) prevent and prepare for chemical emergencies; (2) respond to environmental crises; and (3) inform the public about chemical hazards in their community.

SUMMARY *CEPPO* works with numerous federal agencies, state and local governments and tribes as well as industry, environmental, labor and community groups to reduce risks posed by chemicals in their communities. Until recently, detailed information about the location of hazardous chemicals was accessible, but is now limited because of domestic security concerns. The Web site links to general information and subject-specific data about the Emergency Planning and Community Right-to-Know Act (EPCRA), the risk management planning requirements of the Clean Air Act, up-to-date information on chemical accidents, as well as publications, regulations, conference listings and other databases related to chemical emergency preparedness and accident prevention.

MAIN RESOURCES OFFERED R2, R6, R7, R14

TECHNOLOGY FUNCTIONS T1, T2, T5

SECTOR RELEVANCE various, including chemicals, labor unions, nonprofit organizations, state and local government

CONTACT

E-mail: homepage.ceppo@epamail.epa.gov

RCRA, Superfund & EPCRA Call Center
Tel: 800-424-9346 or 703 412-9810

Technology Innovation Office (TIO)

WEB SITE www.epa.gov/tio/

KEY WORDS Brownfields, groundwater, site characterization, soil, underground storage tanks

PURPOSE to promote more effective and less costly approaches to assessing and cleaning contaminated waste sites, soil and ground water by government and industry.

SUMMARY *TIO*, housed within the Superfund program, works with federal agencies, states, engineering firms, parties responsible for contaminated sites, technology vendors, professional associations, universities and the investment community to enhance the state of site characterization and remediation. TIO documents cost and performance information on new technologies and produces marketing studies for remediation services. The link to REACHIT (Remediation and Characterization Innovative Technologies) provides: 1) a database on technologies searchable by contaminant medium; 2) information about service providers; and 3) sites at which a particular type of technology has been implemented. The *Vendor Support* page, linked from the *Clu-In Database* directs technology developers to funding sources and technical support for all stages of product development and commercialization. The *Partnerships* link displays various projects, which EPA works on with partners within the federal government, such as the *Federal Remediation Technology Roundtable* (see Chapter 4), and outside with public and private entities. *TechDirect*, published monthly, highlights new publications and events of interest. Also posted are guides, newsletters, monographs, technical briefs, brochures, trade journal articles, a vendor list and conference presentations. For example: The *Road Map to Understand Innovative Technology Options for Brownfields Investigation and Cleanup* and the *Vendor Information System for Innovative Treatment Technologies.*

MAIN RESOURCES OFFERED R1, R2, R3, R4, R5, R9, R11, R12, R13, R14

TECHNOLOGY FUNCTIONS T1, T10

SECTOR RELEVANCE chemicals, environmental services, petroleum and professional associations

CONTACTS

Director: Walter W. Kovalick, Jr, Ph.D.
Tel: 703-603-9910
E-mail: kovalick.walter@epa.gov

Vendor Support Page
U.S. EPA (5102G)
Ariel Rios Building
1200 Pennsylvania Avenue, N.W.
Washington, DC 20460
Fax: 703-603-9135

Clu-in Database (must click on vendor support)
www.clu-in.org/remed1.cfm#tech_vend

Other contacts

www.epa.gov/swertio1/about.htm

Sample Technology: Bioventing: In Situ *Biological Treatment for Soil, Sediments and Sludge*

WEB SITE www.epa.gov/swertio1/remed.htm

KEY WORDS air sparging, bioventing, hydrocarbons, pesticide, petroleum, sediment, sludge, soil, solvent, wood preservative.

ENVIRONMENTAL CHALLENGE Traditional technologies (e.g., removal and treatment offsite) often make cleanup of waste sites across the country expensive and time consuming.

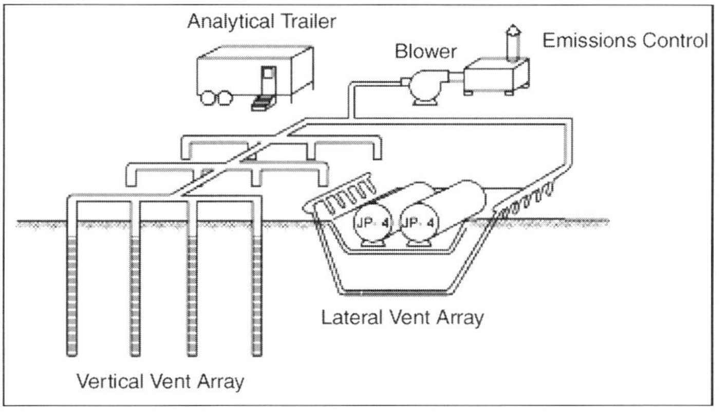

FIGURE 6.12
Typical bioventing system (EPA, with permission).

DESCRIPTION *In situ* bioventing stimulates the natural *in situ* biodegradation of any aerobically degradable compounds in soil by providing oxygen to existing soil microorganisms. In contrast to traditional soil vapor vacuum extraction, bioventing uses low airflow rates to provide only enough oxygen to sustain microbial activity. Oxygen is most commonly supplied through direct air injection into residual contamination in soil. In addition to degradation of adsorbed fuel residuals, volatile compounds are biodegraded as vapors move slowly through biologically active soil. Bioventing is a medium- to long-term technology. Cleanup ranges from a few months to several years.

RESULTS Bioventing techniques have been successfully used to remediate soils contaminated by petroleum hydrocarbons, nonchlorinated solvents, some pesticides, wood preservatives and other organic chemicals. Bioventing is effective under widely varying site conditions. Initial testing has been completed at 117 sites, with more than 90 pilot systems now operating. On smaller sites, many of the single-well pilot systems are providing full-scale remediation. The technology has received regulatory approval in 30 states and all 10 EPA regions and its use in the private sector is growing rapidly.

CONTACT

www.epa.gov/swertio1/about.htm

Underground Storage Tanks (OUST)

WEB SITE www.epa.gov/oust/

KEY WORDS bioventing, groundwater, *in situ*, land use, natural attenuation, petroleum, solvents, surfactant, vapor extraction

PURPOSE to develop and implement a new regulatory program for remediating underground petroleum storage tank leaks.

FIGURE 6.13
Besides removal and replacement, other remedies are available for contamination from underground storage tanks (courtesy Hal White/ OUST).

SUMMARY The *OUST* Web site offers a list of alternative and innovative technologies used to remedy contamination caused by leaking underground tanks. A guidance manual with detailed descriptions of 10 alternative technologies is available for ordering. Online documents describe sampling, monitoring, analysis, site development and contaminants. Also see link to EPA's Environmental Technology Verification (ETV) program (described elsewhere in this chapter).

MAIN RESOURCES OFFERED R1, R12

TECHNOLOGY FUNCTIONS T1, T10

SECTOR RELEVANCE petroleum

CONTACTS

Hal White
U.S. EPA/OSWER/OUST
Tel: 703-603-9900;
Fax: 703-603-0175 or 703-603-9163
E-mail: white.hal@epa.gov
OUST specialists: www.epa.gov/swerust1/oustcont.htm

WasteWise

WEB SITE www.epa.gov/epaoswer/non-hw/reduce/wstewise/

KEY WORDS conservation, industrial process, industrial waste, plastics, recycle/reuse, solid waste

PURPOSE to foster, through partnerships, the reduction of solid waste in industry, small business, nonprofit organizations, institutions and other entities, benefiting their bottom line and the environment.

SUMMARY *WasteWise* provides free technical assistance to more than 1100 volunteer partners to develop, implement and measure their waste-reduction activities. EPA publicizes organizations that are successful in reducing waste through case studies and national and regional events, which provide networking opportunities. Participants sign on for a 3-year period and are required to file annual reports on their progress. Registration occurs on the Web site. A toll-free help line provides general information about waste reduction as well as specific technical queries. The Web site has a searchable directory of reduction resources, including publications and postings from other agencies and trade associations. Upon joining the program, each organization is assigned a representative to provide individual assistance. The publications page displays the annual reports of the program and other publications, such as *How to Start or Expand a Recycling Collection Program* and *Donating Leftover Food to the Needy*. The WasteWise Update is posted periodically. Links to other Web sites identify companies that manufacture recycled products.

Sample Results: (as reported by the companies involved)

Lucent Technologies saved 475,000 lbs of paper and more than $200,000 by increasing the use of e-commerce and reducing the amount of interoffice paper communications.

Guardian Industries cleaned and reused wiping cloths and gloves, reducing waste by 12,000 lbs, saving almost $30,000.

Constellation Energy Group collected 107 million pounds of recyclable materials in 1999, earning the company $1.3 million in sales of recyclables and coal ash products and saving about $360,000.

MAIN RESOURCES OFFERED R1, R2, R4, R5, R10, R12, R14, R15

TECHNOLOGY FUNCTIONS T4, T8, T11

SECTOR RELEVANCE All industries

CONTACT

Toll free helpline: 1800-EPA-WISE (372-9473)
E-mail the WasteWise Helpline at ww@cais.net
WasteWise Program (5306W)
U.S. Environmental Protection Agency
Ariel Rios Building
1200 Pennsylvania Avenue, N.W.
Washington, DC 20460
To register, fill out the form at www.epa.gov/wastewise/about/reg.htm

Office of Wastewater Management (OWM)

FIGURE 6.14
EPA assists wastewater management for small communities.

WEB SITE www.epa.gov/owm/

KEY WORDS wastewater management, wastewater treatment

PURPOSE to assist small communities comply with the wastewater requirements of the Clean Water Act

SUMMARY *OWM* partners with organizations and states to provide technical assistance, financial assistance and education and training for Native American, small and rural communities and rural areas. *Hardship grants* are available to rural areas (population <3000). *State Revolving Funds* (SRFs) provide low-interest loans for a wide variety of water quality projects. Projects can focus on wastewater treatment plants, alternative treatment technologies, estuaries and agricultural, rural and urban runoff. The *National Environmental Training Center for Small Communities* (NETCSC) assists environmental trainers to improve drinking water, wastewater and solid waste services in small communities (population <10,000). The *National Small Flows Clearinghouse* (NSFC) lists manufacturers, consultants and facilities relating to conventional and alternative treatment systems. Also listed are print and audiovisual materials on small community wastewater treatment. The link to EPA's *Onsite/Decentralized Wastewater Systems* site leads to information on technologies, management, funding and public outreach for decentralized wastewater systems.

MAIN RESOURCES AVAILABLE R1, R2, R6, R10, R11, R12, R14

TECHNOLOGY FUNCTIONS T12

SECTOR RELEVANCE Native American tribes, state and local governments, water treatment authorities

CONTACTS

EPA's Office of Wastewater Management
www.epa.gov/owm/contactowm.htm
E-mail: comments.web@epa.gov
For financial assistance:
www.epa.gov/owm/finan.htm#whatwedo
Hardship Grants Program for Rural Communities:

Clean Water State Revolving Fund Program:
U.S. Environmental Protection Agency:
Mailcode 4204
Washington, DC 20460
Tel: 202-260-2268
Fax: 202-260-1827

Clean Water Indian Set-Aside (ISA) Grant Program:
Sylvia Bell
Municipal Support Division (4202)
Office of Wastewater Management
401 M Street, S.W.
Washington, DC 20460
Tel: 202-260-7255
Fax: 202-260-0116
E-mail: bell.sylvia@epamail.epa.gov

NETCSC and NSFC assistance:
Tel: 800-624-8301 or 304-293-4191

Office of Wetlands, Oceans and Watersheds (OWOW)

FIGURE 6.15
A watershed cleanup project (Office of Water, with permission).

WEB SITE www.epa.gov/owow/ www.epa.gov/owow/watershed/outreach/outreach-nonjs.html

KEY WORDS air monitoring, ecosystem, estuaries, groundwater, stormwater, wastewater treatment, water quality, watershed

PURPOSE to promote the management, protection and restoration of water resources and aquatic ecosystems at the watershed level.

SUMMARY *OWOW* provides technical and financial support to states, tribes, local communities and estuary programs and projects. The *Catalog of Federal Funding Sources for Watershed Protection* lists all federal grants and loans searchable by subject. The Oceans and Coastal Protection division of OWOW funds 5-year *National Estuary Programs* in which coastal air deposition monitoring sites are established. The *Five Star Restoration Grant Program* provides an average of $10,000 per project and technical support for projects involving multiple partners among citizen groups, corporations, students, landowners and government agencies. Also available are loans through the *State Revolving Fund* (see description under Office of Wastewater Management). *A Watershed Tools Directory* summarizes several hundred methods, models and data sources that states and communities can use. An *Index of Watershed Indicators* provides data on the condition of rivers, lakes, streams, wetlands and coastal areas. Also see link to the *Watershed Information Resource System Database* (WIRS), a resource center for information on lake and watershed restoration, protection and management.

MAIN RESOURCES OFFERED R1, R2, R4, R6, R12, R14, R15 (economic assessment)

TECHNOLOGY FUNCTIONS T1, T2, T3, T12

SECTOR RELEVANCE citizen groups, communities, Native American tribes, nonprofit organizations, state and local government

CONTACTS

Catalog of Federal Funding Sources for Watershed Protection (Second edition) (include document number in requests: EPA 841-B-99-003)

Tel: 1-800-490-9198

Watershed Tools Directory:
Chris Laabs
Tel: 202-260-7030
E-mail: laabs.chris@epamail.epa.gov

EPA Wetlands Protection Hotline:
Tel: 1-800-832-7828

Wetlands publications:
Tel: 1-800-832-7828
E-mail: wetlands.helpline@epamail.epa.gov.

National Estuaries Program
Tamara Saltman
Tel: 202-260-1459
E-mail: saltman.tamara@epa.gov.

Five Star Restoration Grant Program
John Pai
Tel: 202-260-8076

Sample Technology: Water Restoration on Donald Bair Farm

WEB SITE yosemite1.epa.gov/water/restorat.nsf/Pennsylvania?OpenView

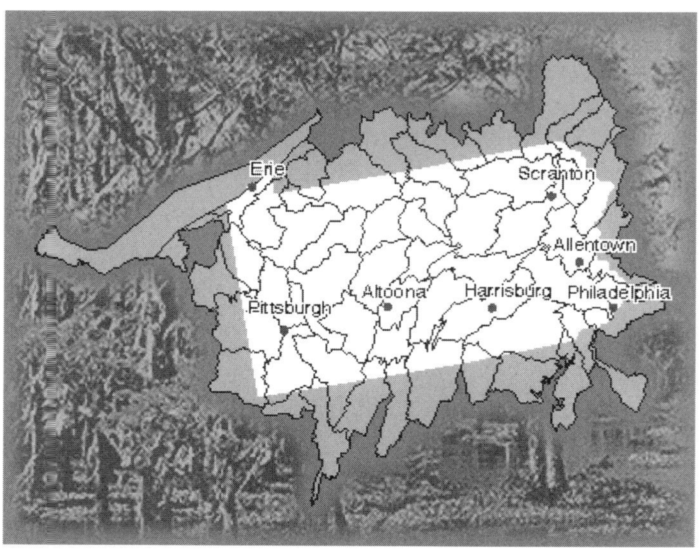

FIGURE 6.16
Water restoration programs in Pennsylvania aim to reduce erosion and sedimentation.

KEY WORDS buffers, erosion, inland waterways, invasive species, nonpoint pollution source, riparian habitat, sediment

ENVIRONMENTAL CHALLENGE Soil, fertilizers and pesticides as well as cattle, dairy, poultry and pig wastes are major pollutants in inland waterways. When deposited into rivers and creeks they cause the rapid growth of invasive species and algae, degradation of habitats and erosion of riverbanks.

PERFORMER Donald Bair Farm in Hereford, PA (90 miles east of Harrisburg). The work involved the Perkiomen Valley Chapter of Trout Unlimited, Schuylkill Riverkeeper, the Patrick Center for Environmental Research, U.S. Fish &Wildlife Service and the Pennsylvania Department of Environmental Protection.

DESCRIPTION A multiphase effort was performed on a 3/4-mile stretch of the farm along a currently unnamed tributary of the Perkiomen Creek. The riparian buffer along the tributary was enhanced by removing the invasive species *multiflora rose,* installing fencing and cattle crossings, installing coir fiber mats and logs to reduce erosion and reforesting the entire stream reach to lessen surface runoff. The project cost approximately $12,000, including bioengineering and plant material, fencing and cattle crossing.

RESULTS This project is currently serving as a demonstration to local farmers. Its success has led to the addition of at least one other fencing or reforestation project on the stream. This, along with a similar past project, will help protect most of the stream.

CONTACT

Schuylkill Riverkeeper
Donald Bair Farm
PO Box 459
St. Peters, PA 19470
Tel: 610-469-6005
Fax: 610-469-6025
E-mail: srk@worldlynx.net

Transportation Air Quality Center (TRAQ)

EPA Office of Transportation and Air Quality

WEB SITE www.epa.gov/otaq/transp.htm

KEY WORDS air quality, greenhouse gases, ozone, public education and outreach, sustainable development, transportation planning

PURPOSE The Transportation Air Quality Center (TRAQ) supports efforts made to protect and improve public health and the environment through integrating air quality and transportation programs.

SUMMARY *TRAQ* provides state and multi-state local air quality regulators, transportation planners and schools with access to information on transportation incentive-based programs, partnership opportunities, grant funding sources and technical assistance. The *Market Incentives ResourceCenter (MIRC)* provides information on programs that use the market, rather than regulatory directives, to encourage reduction of polluting activities or increase the use of less-polluting alternatives. TRAQ offers the COMMUTER Model, usable by people with basic computer skills, which calculates the transportation and emissions benefits of various voluntary strategies to reduce single-occupancy commuting. The TRAQ Web site has a Transportation-Related Grants database that provides information on EPA-funded transportation-related projects. State, local and tribal management agencies, schools and organizations are eligible. The grants address topics such as market incentives, voluntary measures, sustainable development, public education and outreach, transportation control measures and ozone action programs. The TRAQ homepage flashes notices when grant proposals are open.

MAIN RESOURCES R1, R2, R4, R6, R7, R10, R11, R12, R14

TECHNOLOGY FUNCTIONS T2

SECTOR RELEVANCE air quality services, nonprofits, schools, state and local governments, transportation, vehicle manufacturers

CONTACTS

For TRAQ publications:
U.S. Environmental Protection Agency NSCEP
P.O. Box 42419
Cincinnati, OH 45242
Tel: 800-490-9198
Fax: 513-489-8695
Information Request Hotline: 734-214-4100

7

Other Federal Agencies

United States Agency for International Development (USAID)

WEB SITE www.usaid.gov

KEY WORDS biodiversity, climate change, energy efficiency, environmental services, global environmental stewardship, natural resources, renewable energy

PROGRAMS

- Eurasian-American Partnership for Environmentally Sustainable Economies (Ecolinks)
- United States–Asian Environmental Partnership (US-AEP)
  - Environmental Technology Fund
  - Export Advisor
- East Asia and Pacific Environmental Initiative (EA-PEI)
- Office of Private and Voluntary Cooperation (PVC)

PURPOSE to promote United States foreign policy objectives by supporting economic growth, global health and humanitarian assistance to countries in need, including the promotion of environmentally sound technologies and practices.

SUMMARY For U.S.-owned environment firms interested in expanding into international markets, *USAID* matches developing countries' needs with U.S. companies selling appropriate technologies, products and expertise. USAID programs are open to every type of U.S. entity participating in the environmental arena, from environmental and energy businesses to government and multilateral agencies, nonprofits and academia. The 2001 Annual Performance Plan (available in PDF form) breaks down USAID's environmental programs into six focus areas: 1) protecting the world's environment for long-term sustainability; 2) conserving biologically significant habitats; 3) reducing the threat of global climate change; 4) improving the urban populations' access to adequate environmental services; 5) increasing environmentally sound energy services; and 6) promoting sustainable natural resource management. Descriptions of projects within each focus area can be found at www.usaid.gov/environment/whatwedo.html. This site also links to thousands of publications on various programs and projects led by USAID.

MAIN RESOURCES OFFERED R1, R2, R3, R5, R6, R9, R10, R12, R13, R14, R15

TECHNOLOGY FUNCTIONS T1, T2, T3, T4, T6, T7, T8, T9, T13

SECTOR RELEVANCE academia, businesses, indigenous organizations, international governments, international or multilateral agencies, nonprofits, private voluntary organizations

CONTACT

U.S. Agency for International Development Information Center
Ronald Reagan Building
Washington, D.C. 20523-1000
Tel.: 202-712-4810
Fax: 202-216-3524

EcoLinks (USAID)

WEB SITE www.ecolinks.org/

KEY WORDS industrial processes, urban planning

PURPOSE to promote sustainable relationships among businesses, local governments and their associations in Central and Eastern Europe and Eurasia with their U.S. counterparts to mitigate regional urban and industrial environmental problems.

SUMMARY *EcoLinks* fosters development and implementation of practical market-based solutions to industrial and urban environmental problems. EcoLinks pursues these objectives using two complementary mechanisms: the *Partnership Grants Program* and the *Trade and Investment Program*. EcoLinks provides grants to foster partnerships involving businesses, local governments and their associates. *Challenge Grants* provide up to $50,000 for 1-year partnerships. *Quick Response Awards* of up to $5,000 are for the immediate and small-scale start-up needs of potential partnerships within EcoLinks. The EcoLinks Grants are implemented through three regional offices in Szentendre, Hungary, Washington, D.C. and Kiev, Ukraine. Seven local offices are situated in Eastern Europe. The Web site has a downloadable application form, the grants issued during the last 4 fiscal years and an index of topics with links to information relevant to the partnerships, such as urban management and sustainable development. The Trade and Investment Program facilitates trade and investment in environmental goods and services. In addition, EcoLinks provides information about environmental laws, policies and regulations, environmental technologies, and management systems in the target countries. EcoLinks also supports the *U.S. Clean Technology Exchange*—a free forum for information exchange on environmental technologies and project implementation. Access to the technologies offered by members requires registration and profile creation.

MAIN RESOURCES OFFERED R2, R4, R5, R6, R10, R12, R15

TECHNOLOGY FUNCTIONS T1, T6, T13

SECTOR RELEVANCE businesses, emerging-market nations, local governments, various

CONTACT

Partnership Grants Program
Mr. Gerald Gold
Tel.: 202-712-0263
Fax: 202-216-3014
E-mail: ggold@usaid.gov

Trade and Investment Program
Mr. Jeffrey Orrey
Tel.: 202-712-1608
Fax: 202-216-3014
E-mail: jorrey@usaid.gov

United States–Asia Environmental Partnership (US-AEP) — USAID

WEB SITE www.usaep.org

KEY WORDS energy efficiency, greenhouse gases, industrial processes, pollution prevention

PURPOSE to promote environmentally sustainable development in Asia through partnerships between U.S. and Asian governments, NGOs, academia and the private sector.

SUMMARY *US-AEP* is a public–private initiative that works with governments and industries in Hong Kong, India, Indonesia, Korea, Malaysia, Philippines, Singapore, Sri Lanka, Taiwan, Thailand and Vietnam in four program areas: policy, urban, industry and technology cooperation. The *Environmental Technology Network for Asia (ETNA)*, in conjunction with the *Global Technology Network (GTN)*, is a U.S.-government-sponsored clearinghouse that collects information on trade opportunities in Asia, forwards it to U.S. environmental companies and provides counseling and market-trend analysis on Asian countries. U.S.-owned environmental companies are eligible to receive trade leads after registering online. US-AEP's *Clean Technology and Environmental Management* supports initiatives such as *Greening the Supply Chain* and *Voluntary Business Standards* to promote industrial environmental performance. The monthly *US-AEP Update* has information on conferences, current projects and new initiatives. The Reports and Publications link leads to case studies, new government policies and future plans. The *Export Advisor* offers country comparisons and help in identifying prospective markets.

Sample Resource: Environmental Technology Fund

The Environmental Technology Fund provides grants to small to medium-sized (with less than $50 million in annual revenues) U.S. companies. They must be sponsored, however, by a nonprofit organization, such as a state development agency, trade association, chamber of commerce, or other intermediary. Projects may address pollution control, pollution prevention or clean technology, and energy efficiency in priority Asian economies. Preferred projects aim to reduce or eliminate pollution at the source and reduce emissions of global greenhouse gases. Funding of up to $20,000 covers 20–50% of a project's total cost. Eligible activities include engineering and technology workshops or seminars, business development missions and technology and equipment demonstrations.

MAIN RESOURCES OFFERED R1, R3, R4, R5, R6, R9, R10, R11, R14, R15

TECHNOLOGY FUNCTIONS T1, T4, T5, T6, T13

SECTOR RELEVANCE various, including industry, nonprofits, small to medium-sized businesses, state and local government

CONTACTS

US-AEP, Asia/Near East Bureau
U.S. Agency for International Development
1300 Pennsylvania Avenue, NW, 4th Floor
Washington, DC 20523-4101, USA

Peter Kimm, Executive Director
Tel.: 202-712-4156
Fax: 202-216-3379
E-mail: pekimm@usaid.gov

For Environmental Technology Fund information:
National Association of State Development Agencies
Attn: Environmental Technology Fund
750 First St NE, Suite 710
Washington, DC 20002, USA
Tel.: 202-326-6253
Fax: 202-898-1312

Office of Private and Voluntary Cooperation (PVC) — (USAID)

WEB SITE www.usaid.gov/hum_response/pvc/index.html

KEY WORDS sustainable development.

Purpose to strengthen the capacity of local nongovernmental organizations (NGOs) and community groups to deliver sustainable services in developing countries and emerging democracies.

SUMMARY *PVC* provides direct support to efforts made by U.S. private and voluntary organizations (PVOs) and NGOs in five USAID strategic areas, including The World's Environment Protected for Long-Term Sustainability. Through the *Matching Grants Program*, PVOs registered with USAID can replicate successful programs in new places or initiate new projects. Awards are based on demonstrated ability to implement successful sustainable development programs. *Cooperative Development Grants* encourage the establishment of long-term partnerships between U.S. Cooperative Development Organizations (CDOs) and host countries overseas. Applicants must be a U.S. cooperative, a CDO, or an association representing cooperative members. The publications link has information on current programs, current country statistics, a brochure of USAID resources available to PVOs and contact information for registration.

MAIN RESOURCES OFFERED R1, R4, R5, R6, R9, R10, R14, R15

TECHNOLOGY FUNCTIONS T6, T13

SECTOR RELEVANCE cooperatives, NGOs, private and voluntary organizations

CONTACTS

Matching Grants Program:
Martin Hewitt, Team Leader, Matching Grant Program
Tel.: 202-712-1531
Fax: 202-216-3041
Email: mhewitt@usaid.gov

Cooperative Development Grants Program:
Tom Carter
Technical Advisor, Cooperative Development Program
Tel.: 202-712-5226
Fax: 202-216-3039
Email: thcarter@usaid.gov

United States Department of Agriculture (USDA)

Sustainable Agriculture Research and Education Program (SARE)

FIGURE 7.1
SARE helps ranchers and farmers adopt practices that are environmentally sound and economically viable ((USDA–SARE, with permission).

Web Site www.sare.org

KEY WORDS sustainable agriculture

PURPOSE to integrate agricultural practices with the surrounding environment in order to maintain a healthy environment while still producing high-quality products.

SUMMARY *SARE* has four regional councils (Northeast, South, North Central and West) that recommend projects to be funded. Since 1988, SARE has funded more than 1600 projects through its three competitive grants programs. Research and education grants generally range from $30,000–$200,000. The projects usually involve scientists, producers and other parties in an interdisciplinary approach. Professional development grants, generally in one state—often on one farm—range up to $5000 or $10,000, depending on the region. Producer grants typically range from $500 to $15,000 to conduct onsite research experiments and share the results with other farmers and ranchers. Applications for grants must go through the regional offices. The Web site contains application and deadline information. The SARE 2000 highlights page (www.sare.org/2000high/intro.htm) links to descriptions of recently funded projects.

Sample Technology: Organic Farming

Full Circle Solutions, Inc, with the help of a SARE grant, designed a system for small farms that converts food waste to liquid fertilizer as an alternative to expensive organic fertilizers. Called "the digester," the device resides at Possum Hollow Farm, where the owner recycles approximately 2 tons of cafeteria waste weekly. Anaerobic bacteria in the digester convert the food waste to methane and carbon dioxide, while the nutrients remain in the water or in a small amount of solid residue that can be land applied or cured to mature compost. The fertilizer nitrogen value produced from 2 tons of food waste per week ranges from $800 to more than $10,000 annually. Additional farm income could be derived from tipping fees for accepting food waste; a $20-per-ton tipping fee would produce $2,000 extra annual income.

MAIN RESOURCES OFFERED R1, R3, R6, R12

TECHNOLOGY FUNCTIONS T4, T6, T7, T8, T9, T13

SECTOR RELEVANCE agriculture, state and local governments.

CONTACTS

General Information about Grant Procedures:

Tel: 202-720-5203

E-Mail: ehauhn@reeusda.gov

Media and Communications:

Tel: 301-405-3186;

E-mail: vberton@wam.umd.edu

Regional Information:

www.sare.org/2000high/about.htm

To join the electronic listserv "sanet-mg," send a message to:

listserv@sare.org. In the body of the message, type "subscribe sanet-mg" and your full name.

United States Department of the Interior (DOI)

Clean Streams Initiative (Office of Surface Mining)

WEB SITE www.osmre.gov//acsihome.htm

KEY WORDS mine tailings, river basin

PURPOSE to facilitate the efforts of citizen groups, university researchers, the coal industry, corporations, the environmental community and local, state and federal government agencies to clean streams polluted by mine drainage.

SUMMARY *OSM* funds the *Clean Streams Initiative Program* through grants, cooperative agreements and an internship program. Through its *Abandoned Mine Land (AML) Grant Program*, states or tribes located near coal mining operations that pay coal reclamation fees are eligible for grants ranging from \$100,000 to \$28 million. The amount is based on historical coal production and current fee collections. The *Watershed Cooperative Agreement Program* provides funds of up to \$100,000 to local nonprofit groups with projects that involve construction of infrastructure to eliminate pollution discharges into streams. To receive funding, applicants must gain additional funding from other partners. The *Summer Watershed Internship Program* places interns with the local watershed groups. OSM provides each intern a stipend of \$2500. The Clean Streams Program Web site has information on potential funding opportunities, assistance for grant writing, eligibility requirements, the application process and an application form.

MAIN RESOURCES OFFERED R1, R5, R6, R10, R12, R15

TECHNOLOGY FUNCTIONS T1, T3, T10, T13

SECTOR RELEVANCE academia, mining, nonprofits, federal, state and local governments

CONTACTS

Appalachian Clean Streams Initiative
Office of Surface Mining
1951 Constitution Ave NW
Washington, DC 20240-0003
Milton Allen
Tel.: 412-937-2863
E-mail: mallen@osmre.gov

For a list of state contacts:
www.osmre.gov/acsicontacts.htm

To subscribe to OSM's list-serve, e-mail majordomo@osmre.gov. In the body of the message type "subscribe clean stream" and your Internet address.

National Coastal Wetlands Conservation Grants Program (U.S. Fish and Wildlife Service)

WEB SITE www.fws.gov/cep/coastweb.html

KEY WORDS coastal wetlands

PURPOSE to protect coastal wetlands from the threats of agricultural and urban run-off, shoreline modification, invasive species, oil spills, residential and commercial development, municipal waste disposal and over-harvesting of resources.

SUMMARY The *National Coastal Wetlands Program* provides matching grants for acquisition, restoration, management, or enhancement of coastal wetlands. The program benefits have included habitat protection and restoration for migratory birds, shorebirds, waterfowl, certain insects, and fish and shellfish. Between $11 million and $15 million in grants are awarded annually. Awards cannot exceed $1 million for an individual project. Those eligible for monies are states that border the Atlantic, the Pacific, the Gulf of Mexico and the Great Lakes as well as the trust territories and commonwealths of the United States. Excluded is Louisiana, which has its own coastal wetlands program. States must provide at least 50% of the total cost of the project. The federal share can be increased to 75% if the state establishes and maintains a special fund for acquiring coastal wetlands. Puerto Rico is the only territory or commonwealth required to share the project cost. To date, grants totaling about $90 million have been awarded to 25 coastal states and one U.S. territory. Project descriptions for FY2000 and 2001, as well as breakdowns of funding, can be found in the news release section of the U.S. Fish and Wildlife Service (FWS) Web site. Virtually all other information, including proposal requirements, is located in the Federal Register, Vol. 66, No. 161, pp. 43555-67.

MAIN RESOURCES OFFERED R1, R6

TECHNOLOGY FUNCTIONS T3

SECTOR RELEVANCE state and local governments, U.S. Territories

CONTACTS

Division of Fish and Wildlife Management
Branch of Habitat Restoration
Tel.: 703-358-2201
Migratory Birds and State Programs
Division of Federal Aid
Tel.: 703-358-2156

For News Releases:
Rachel Levin
Office of Public Affairs
Washington, DC 20240
Tel.: 202-208-5634
Fax: 202-219-2428

United States Geological Survey (USGS)

WEB SITE www.usgs.gov

KEY WORDS monitoring, remediation

PURPOSE to provide reliable scientific information to manage water, biological, energy and mineral resources.

SUMMARY *USGS* awards grants, cooperative agreements and contracts to manage the nation's resources. The *Technology Enterprise Office* links the research organizations within USGS to the private-sector parties interested in accessing USGS technology through patent licensing and cooperative research and development agreements (CRADAs). Research and technology transfer opportunities are available in the following general areas: environmental cleanup; resource detection and assessment; sampling technologies for water, air and soil; modeling and prediction software, information and data management; visualization and mapping; educational materials and biotechnology. The USGS has technologies available for licensing and lists contacts within DOI for its technology transfer programs.

MAIN RESOURCES OFFERED R6, R8, R12

TECHNOLOGY FUNCTIONS T1, T3, T13

SECTOR RELEVANCE construction, fuels, mining

CONTACTS

Charles G. Groat, Director
U.S. Geological Survey
Headquarters, John W. Powell Federal Building
12201 Sunrise Valley Drive
Reston, VA 20192
Tel.: 703-648-7411

For Grant information:
Ric Heilemann
Business Utilization and Development Specialist
U.S. Geological Survey
12201 Sunrise Valley Drive
Mail Stop 205P
Reston, VA 20192
Tel.: 703-648-7346
E-mail: rheilema@usgs.gov

For grant eligibility requirements and proposal information:
www.usgs.gov/contracts/grants/

Technology Enterprise Office:
Julia Giller
U.S. Geological Survey
211 National Center
12201 Sunrise Valley Drive
Reston, VA 20192
Tel.: 703-648-4450
E-mail: jgiller@usgs.gov

Department of Transportation (DOT)

Clean Fuels Formula Program

FIGURE 7.2
CFFP provides funds for the development of clean-fuel technologies for urban transportation.

WEB SITE www.fta.dot.gov/research/equip/buseq/cfuels/cfuels.htm

KEY WORDS air quality, clean fuels

PURPOSE to support emerging clean-fuel technologies, to finance their purchase or lease and to support markets for new technologies as well as assist nonattainment and maintenance areas in achieving or maintaining attainment status.

SUMMARY Under the newly authorized *Clean Fuels Formula Program,* which merged with the *Bus Capital Program,* cities in maintenance and nonattainment air-quality areas, as defined by the EPA, can receive funds to acquire clean-fuel vehicles, to re-power or retrofit engines for clean-fuel operation and to construct or improve facilities to support these vehicles. Funds are then distributed by a formula based on size of the bus fleet, bus passenger miles and the severity of non-attainment for ozone and carbon monoxide. Two thirds of the funds are appropriated to cities with populations greater than 1 million, the remainder going to cities with smaller populations.

MAIN RESOURCES OFFERED R6

TECHNOLOGY FUNCTIONS T2, T8

SECTOR RELEVANCE state and local government

CONTACT

The Office of Research, Demonstration and Innovation
Tel.: 202-366-4052
Federal Transit Administration
400 7th Street, SW
Washington, DC 20590
Tel.: 202-366-1635

Export-Import Bank (Ex-Im Bank)

Environmental Exports Program (EEP)

WEB SITE www.exim.gov/envprogs.html

KEY WORDS air quality, ecological assessment, monitoring, renewable energy, toxic substances, water treatment

PURPOSE to support exporters of environmentally beneficial goods and services and exporters participating in foreign environmentally beneficial projects.

SUMMARY *EEP* supports large and small exporters of environmental products and services. The program offers two types of export assistance: 1) a short-term insurance policy providing single- and multi-buyer coverage for small-business environmental exporters and 2) an enhanced loan and guarantee program, including local cost coverage equal to 15% of the U.S. contract price and capitalization of interest during construction. Exports eligible for support must be used for abatement, control or prevention of air, water and ground contamination or pollution, or for protection in the handling of toxic substances. Examples of these are: instruments to measure or monitor air or water quality; emission-control devices; effluent pollution-control devices; equipment for systems for waste disposal, refuse collection and wastewater treatment; services to upgrade environmental regulations; ecological studies; ecological monitoring equipment; toxic material handling devices and certain renewable and alternative-energy equipment. For products to qualify for the insurance policy, they must be manufactured in the United States and have at least 51% U.S. content and service.

MAIN RESOURCES OFFERED R7, R15

TECHNOLOGY FUNCTIONS various

SECTOR RELEVANCE environmental equipment, environmental services

CONTACT

Export-Import Bank of the U.S.
Washington, D.C. Office
811 Vermont Ave., NW
Washington, DC 20571
Tel.: 800-565-EXIM (3946)
Tel.: 202-565-EXIM (3946)
Fax: 202-565-3380

Federal Trade Commission (FTC)

Bureau of Consumer Protection

WEB SITE www.ftc.gov/bcp/conline/edcams/eande/index.html

KEY WORDS energy efficiency, environmental labeling

PURPOSE to provide businesses and consumers with important information about energy and environmental issues.

SUMMARY The Bureau of Consumer Protection's *Energy and Environment* department includes resources on energy and environmental labeling for consumer products; retail sales of electricity, home insulation and home energy and fuel; and automotive products. The appliances site includes the FTC's *Appliance Labeling Rule*, which helps consumers comparison shop for energy-efficient products, including central and room air conditioners, dishwashers, washing machines, pool heaters, furnaces, refrigerators and water heaters. FTC requires manufacturers to attach labels to products indicating their energy consumption and efficiency. The site also provides compliance information for businesses. The environment site includes the FTC's *Environmental Marketing Guides*, or "Green Guides," designed to prevent false or misleading environmental marketing claims and to explain compliance measures, which apply to all forms of marketing — advertisements, labels, package inserts, promotional materials, words, symbols, logos, product brand names and via the Internet or e-mail. The "enforcement" page lists specific claims contested by FTC and "web links" offers texts of labeling rules for appliances and insulation. The fuels and automotive page covers labels on gasoline pumps and alternative fuel vehicles, guides to cleaner fuel, alternative-vehicle labels and recent fuel price comparisons.

MAIN RESOURCES OFFERED R15

TECHNOLOGY FUNCTIONS T4, T13

SECTOR RELEVANCE consumers, manufacturing, petroleum

CONTACT

Federal Trade Commission
CRC-240
Washington, D.C. 20580
Tel.: 202-326-2222
Tel.: 877-FTC-HELP (382-4357)

Internal Revenue Service (IRS)

Tax Credit and Deductions

KEY WORDS clean fuel, energy efficiency, pollution abatement, pollution prevention, renewable energy, underground storage tanks

PURPOSE to provide America's taxpayers with top quality service by helping them understand and meet their tax responsibilities and by applying the tax law with integrity and fairness to all.

SUMMARY Several provisions of the U.S. Tax Code and IRS rulings address environment-related matters (see selected listing below). For advice on applicability of any particular provision of the tax code, check with the IRS or a tax professional.

Section 30—Credit for qualified electric vehicles. In general, purchasers of qualified electric vehicles are allowed a credit equal to 10% of the cost of the vehicle for the taxable year that the vehicle is placed in service. The maximum amount of the credit allowed with respect to any qualified electric vehicle may not exceed $4000.

Section 48—Solar energy and reforestation credit. In general, taxpayers gain a 10% credit of the basis of each energy property. Energy property is 1) any property that uses solar energy to generate electricity, to heat or cool, or to provide process heat and 2) equipment used to produce, distribute, or use energy derived from geothermal deposits. Section 48(b) provides for a 10% credit for certain costs of reforestation.

Section 136—Energy conservation subsidies provided by public utilities. A public utility customer may exclude from gross income the value of any subsidy provided, directly or indirectly, by the utility for the purchase or installation of an energy conservation measure, which is any installation or modification primarily designed to reduce the consumption of electricity or natural gas or to improve the management of energy demand at a dwelling unit. The provision does not apply to any payment to or from a qualified cogeneration facility or qualifying small power production facility pursuant to section 210 of the Public Utility Regulatory policy Act of 1978.

Section 169—Amortization of pollution-control facilities. Provides for an election to amortize the cost of any certified pollution-control facility straight-line over a 60-month period instead of depreciation. A certified pollution-control facility is a new identifiable treatment facility used in connection with a plant or other property in operation before January 1, 1976, to abate or control water or atmospheric pollution or contamination by removing, altering, disposing, storing, or preventing the creation or emission of pollutants, contaminates, wastes, or heat. State and federal authorities must certify the facility to be in conformity with the state program or requirements and to be in furtherance of the general policy under the Federal Water Control Act or the Clean Air Act.

Section 175—Soil- and water-conservation expenditures. A taxpayer engaged in the business of farming may elect to treat certain soil- and water-conservation expenditures with respect to land used in farming as expenses. Under this election, conservation expenditures that would otherwise be classified as capital expenditures under section 263 may be deducted currently subject to certain requirements and limitations. The election does not apply unless the expenditures are consistent with the appropriate federal or state conservation programs. The deduction is limited annually to 25% of the taxpayer's income from farming. Any excess expenditures may be carried forward and deducted in succeeding taxable years.

Section 179A—Deduction for clean-fuel vehicles and certain refueling property (expires for property placed in service after December 31, 2006. In general, taxpayers may deduct 1) the costs of qualified clean-fuel vehicle property for the taxable year that the property is placed in service and 2) the cost of qualified clean-fuel vehicle refueling property for the taxable year that the property is placed in service. The amount of the deduction for clean-fuel vehicle property is based on the size of the motor vehicle. In general, $2000 may be deducted for a qualified automobile. For qualified clean-fuel vehicle refueling property placed in service during the taxable year at a location, the deduction may not exceed the excess (if any) of $100,000, over the aggregate amount of section 179A deductions taken by the taxpayer with respect to such property placed in service at such location for all preceding taxable years.

Section 198—Expensing of environmental remediation costs. Generally permits taxpayers to elect to treat any qualified environmental remediation expenditure as a deduction for the taxable year that such expense is paid or incurred. Qualified environmental remediation expenditures include any expenditure that is otherwise chargeable to a capital account and that is paid or incurred in connection with the abatement or control of hazardous substances at a qualified contaminated site. Qualified contaminated sites are certain targeted areas that are held by a taxpayer for use in its trade or business, for the production of income, or as inventory, and upon which there has been a release (or threat of release) or disposal of any hazardous substance. These targeted areas include 1) any population census tract with a poverty rate of not less than 20%, 2) any population census tract with a population of less than 2000 if more than 75% of such tract is zoned for commercial and industrial use and such tract is contiguous to one or more other population census tracts with a poverty rate of not less than 20%, 3) any empowerment zone or enterprise community (and any supplemental zone designated on December 21, 1994) and 4) any site announced before February 1, 1997, as being included as a Brownfields pilot project of the Environmental Protection Agency.

Rev. Rul. 94-38, 1994-1 C.B. 35—Groundwater contamination. IRS holds that costs to clean up land and to treat groundwater that a taxpayer contaminated with hazardous waste from its business are deductible by the taxpayer as ordinary and necessary business expenses under section 162. The service also holds that costs allocable to the construction of a groundwater treatment facility must be treated as capital expenditures under section 263 and are recoverable by the taxpayer under the applicable cost recovery systems (e.g., section 168).

Rev. Rul. 98-25, 1998-19 I.R.B. 4—Underground storage tanks. IRS holds that costs incurred to replace underground storage tanks containing waste by-products (including the cost of removing, cleaning and disposing of the old storage tanks and acquiring, installing and filling the new storage tanks) are deductible by the taxpayer as ordinary and necessary business expenses under section 162. The revenue ruling also notes that the taxpayer would be permitted to deduct the costs of removing and replacing above-ground storage tanks in the same context.

MAIN RESOURCES OFFERED R7

TECHNOLOGY FUNCTIONS T2, T4, T8, T10, T12

SECTOR RELEVANCE agriculture, automotive, business, consumer, environmental services, public utility

National Aeronautics and Space Administration (NASA)

Commercial Technology Program

FIGURE 7.3
Caption: NASA licensed hi-tech communications gear to KVH Industries, Inc. for commercial development of improved satellite antenna communication from marine or mobile land transportation, which could be used to transmit environmental data from the field

WEB SITE www.sti.nasa.gov/tto/spinoff1997/contents.html

KEY WORDS technology transfer

PURPOSE to assist U.S. businesses and industry in accessing, utilizing and commercializing NASA-funded research and technology by providing a full range of technology transfer and commercialization services and assistance.

SUMMARY Within the *Commercial Technology Program* are the *National Technology Transfer Center* (NTTC) and six *Regional Technology Transfer Centers* (RTTCs). The NTTC links U.S. companies with federal technological resources and provides businesses with access to NASA and other federal technologies, expertise and facilities. It also administers the *Technology Tracking System* (TechTracS), an online network of databases designed to maintain an inventory of NASA technologies, help manage the identification of new technologies and their commercial potential, administer intellectual-property protection and seek commercialization partners. The six RTTCs are: Northeast, Far West, Mid-Atlantic, Southeast, Midwest and Mid-Continent. Each RTTC has the same purpose as the NTTC, but on a regional level. The environmental and resource section describes various technologies developed by federal labs, including ways of monitoring earth's ecosystems and atmosphere.

Sample Technology: Tire Recycling

Cryopolymers, Inc. of St. Francisville, Louisiana, in conjunction with the Southeast Regional Technology Transfer Center and John C. Stennis Space Center in Mississippi, developed a way to improve the recycling of vehicle tires to convert shredded rubber into products that can be used in asphalt roadbeds, new tires, hoses and other products. Cryopolymers used NASA expertise for effective methods of freezing tires to separate the rubber from the other materials. For each pound of rubber salvaged, nearly 90% is reduced into crumb. Crumb can contribute to new tires, agricultural hoses, or, when mixed with plastics, culvert piping and protective mats for the bed of pickup trucks. What remains is called "fluff," scrap metal and polyester residue that can be incorporated into new products as a reinforcing fiber. Cryopolymers expects to recycle 5000 tires per day, yielding an estimated annual income of $4 million.

MAIN RESOURCES OFFERED R1, R2, R10, R12

TECHNOLOGY FUNCTIONS T13

SECTOR RELEVANCE rubber products

CONTACTS

NASA Systems Division Office
Valerie J. Steele, Division Secretary
Tel.: 650-604-4502
Fax: 650-604-437
E-mail: vsteele@mail.arc.nasa.gov

Bryan Biegel, Research Branch Chief
Tel.: 650-604-0171
E-mail: bbiegel@mail.arc.nasa.gov

National Technology Transfer Center:
www.nttc.edu

Technology Tracking System (TechTracS):
technology.ksc.nasa.gov/TechTracS/TechTracS.html

Marshall Space Flight Center Technology Transfer Program (MSFC)

WEB SITE www.nasasolutions.com/index.html

KEY WORDS technology transfer

PURPOSE to encourage broad use of Marshall-developed technologies by American private enterprise.

SUMMARY *MSFC* partners with existing national, regional, state and local technology transfer networks. It works with companies to integrate technology into a product line and to develop those technologies into new products. MSFC usually formalizes a relationship with a company via a patent or copyright license or a space act agreement, which is similar to a CRADA but more flexible. As part of the process, MSFC hosts a technology briefing for companies' technical or management representatives. An outreach team publicizes those technologies available for transfer, as well as success stories that can be found on the Web site dating back to 1996.

Sample Technology: Metal Coating Technology

In 1997, The Convergent Spray Technologies™ (CST™) spray process, developed by USBI Co., was used by NASA's Marshall Space Flight Center to apply heat-resistant coatings to the space shuttle's solid rocket boosters. A recently completed MSFC/USBI project demonstrated the feasibility of applying the space-age technology to the roofing industry. This was accomplished by producing an environmentally compliant coating that used recycled materials to coat metal roofs. The project culminated in the application of a coating to a 960-square-foot metal roof at the Marshall Center. Tests have shown the coating to have satisfactory moisture resistance, adhesion properties, tensile strength, flexibility, resistance to water vapor penetration and resistance to the cyclic effects of aging, rain, dew and ultraviolet radiation (sunlight) exposure.

MAIN RESOURCES OFFERED R1, R3, R4, R5, R8, R10, R12

TECHNOLOGY FUNCTIONS various

SECTOR RELEVANCE various

CONTACTS

NASA MSFC Space Flight Center
Technology Transfer Department
Mail Code: CD30
Huntsville, AL 35812
Tel.: 256-544-4266
Fax: 256-544-3151
E-mail: ttd.help@msfc.nasa.gov

Outreach Team
Tel.: 256-544-0710
E-mail: ttd.outreach@msfc.nasa.gov
For application requirements and other information for patent acquisition: techtran.msfc.nasa.gov/working/patent.html

For a Space Act Agreement form:
techtran.msfc.nasa.gov/working/crf.pdf

National Science Foundation (NSF)

WEB SITE www.nsf.gov

KEY WORDS ecological assessment, information technology.

PROGRAMS

- Design, Manufacture and Industrial Innovation Division (DMII)
- Manufacturing Process and Equipment Systems
- Technology for a Sustainable Environment (TSE)
- Partnerships for Innovation (PFI)

PURPOSE to partner with academia, industry and state and local governments to advance science, engineering, mathematics and technology through research and education programs.

SUMMARY *NSF* supports, through grants and contracts, nearly 20,000 research and education projects in science and engineering. Some of its partnerships include the *Engineering Research Centers*, which integrate academia and industry to focus on technological advances. One center, the NSF/SRC Engineering Research Center for Environmentally Benign Semiconductor Manufacturing, has a specific environmental focus, but others develop innovative materials and processes that have environmental values. ERC's homepage lists the centers. NSF also supports small business innovation research (SBIR) programs and small business technology transfer (SBTT) programs (see description in Chapter 4 under the Small Business Administration entry). NSF partners with EPA in a *Decision-Making and Valuation for Environmental Policy* grants program. NSF/EPA will award about ten grants within their budget of $1.5 million. The specific research areas for the grants focus on 1) ecosystem valuation 2) research linking individual environmental values with those of communities and 3) the role of environmental information in environmental decision making.

MAIN RESOURCES OFFERED R5, R6, R10, R12

TECHNOLOGY FUNCTIONS T13

SECTOR RELEVANCE academia, industry, state, local and foreign governments

CONTACTS

The National Science Foundation
4201 Wilson Boulevard
Arlington, VA 22230
Tel.: 703-292-5111

ERC Homepage:
www.eng.nsf.gov/eec/erc.htm

General Grant Information:
www.nsf.gov/home/grants.htm

Decision-Making and Valuation for Environmental Policy Grants:

Susan Carrillo
Tel.: 202-564-4664
E-mail: carrillo.susan@epa.gov
es.epa.gov/ncer/rfa/02dmvep.html

Design, Manufacture and Industrial Innovation Division (DMII)

WEB SITE www.eng.nsf.gov/dmii/

KEY WORDS design for the environment, energy efficiency, material substitutes, manufacturing technology

PURPOSE to support and manage technology and engineering crosscutting programs between business and industry in fields with significant market opportunities.

SUMMARY *DMII*-funded research includes environmental considerations in manufacturing processes and materials. The *Materials Processing and Manufacturing Program* focuses on environmentally benign manufacturing, virtual manufacturing and novel hybrid processing—new processes that optimize materials and energy use to achieve products with superior quality and performance. The Web site lists current research areas, upcoming workshops and past workshop reports. Areas not supported by the program include continuous chemical synthesis and processes; material investigations not directly related to processing issues; and semiconductor materials and processing. The main page includes proposal submission information, deadlines, a downloadable tutorial of the process, and conference and workshop information. DMII also supports small business innovation research (SBIR) and small business technology transfer (SBTT).

MAIN RESOURCES OFFERED R4, R6, R10, R12, R15

TECHNOLOGY FUNCTIONS T4, T6, T11, T13

SECTOR RELEVANCE academia, manufacturing industries, small business

CONTACTS

Design, Manufacture and Industrial Innovation Division
4201 Wilson Boulevard, Suite 550
Arlington, VA 22230
Tel.: 703-292-8330
Fax: 703-292-9056
E-mail: dmii@nsf.gov

Materials Processing and Manufacturing Program:
Dr. Delcie R. Durham, Program Director
Tel.: 703-292-7060
E-mail: ddurham@nsf.gov

Technology for a Sustainable Environment (TSE)

WEB SITE www.nsf.gov/home/crssprgm/epa/start/tse.htm

KEY WORDS design for the environment, green chemistry, industrial ecology

PURPOSE to advance the discovery, development and use of innovative technologies to avoid or minimize the generation of pollutant material.

SUMMARY NSF and EPA jointly support *TSE*, which funds research that will lead to environmentally benign methods for industrial processing and manufacturing. Competitive proposals address all manufacturing processes, including design, production and waste removal. Research areas are chemistry and chemical reaction-based engineering for pollution avoidance or prevention; nonreaction-based engineering for pollution avoidance and prevention; and green design, manufacturing and industrial ecology. Specific examples include alternative reaction conditions, safer chemicals, material substitutions and process alternatives, bioengineering and life-cycle assessments (LCA). Academic and nonprofit institutions are eligible for funding. TSE makes approximately 20 awards, which range from $50,000 to $125,000 annually for 1 to 3 years. The main Web site links to tables of previous grants with research project titles and names of principal investigators and their institution.

MAIN RESOURCES OFFERED R4, R6

TECHNOLOGY FUNCTIONS T13

SECTOR RELEVANCE academia, chemicals, manufacturing, nonprofits

CONTACTS

To reach the TSE Web site, go to www.nsf.gov and type "technology for a sustainable environment" in the search field.

NSF/EPA Partnership for Environmental Research
Dr. Robert Wellek
NSF Directorate for Engineering
Fax: 703-292-9054
E-mail: rwellek@nsf.gov

Technology for a Sustainable Environment (TSE)
Dr. Barbara Karn
EPA Environmental Engineering Research Division
Tel.: 202-564-6824
E-mail: karn.barbara@epa.gov

Partnerships for Innovation (PFI)

WEB SITE www.ehr.nsf.gov/pfi/

KEY WORDS aquaculture, fuel cells, materials substitution

PURPOSE to support innovation in communities by developing the infrastructure needed to connect new scientific discoveries to practical uses.

SUMMARY *PFI* partnerships focus on a specific technology, industry, or geographic area. Recent environmental projects include Advanced Materials for Fuel Cell Systems and Advanced Polymer Materials for Construction and Aquaculture Marketing Development. Grants range from about $300,000 to $600,000 for 2- to 3-year projects. Approximately 15–25 grants are awarded annually. Partnerships may be formed between academia, nonprofits, private businesses, venture capitalists, state and local governments and federal laboratories. Degree-granting institutions of higher learning must submit the proposals and lead the partnership. Awardees must share 10% of the requested budget. The Web site provides proposal format and content information, budgetary limitations and contact information.

MAIN RESOURCES OFFERED R1, R6, R10

TECHNOLOGY FUNCTIONS T6

SECTOR RELEVANCE academia, federal laboratories, nonprofits, private businesses, state and local governments, venture capitalists

CONTACT

John Hurt, Program Director
Partnerships for Innovation Program
National Science Foundation
4201 Wilson Blvd, Room 815
Arlington, VA 22153
Tel.: 703-292-5332
Fax: 703-292-9018
E-mail: jhurt@nsf.gov

8

International Agencies

African Development Bank Group (AFDB)

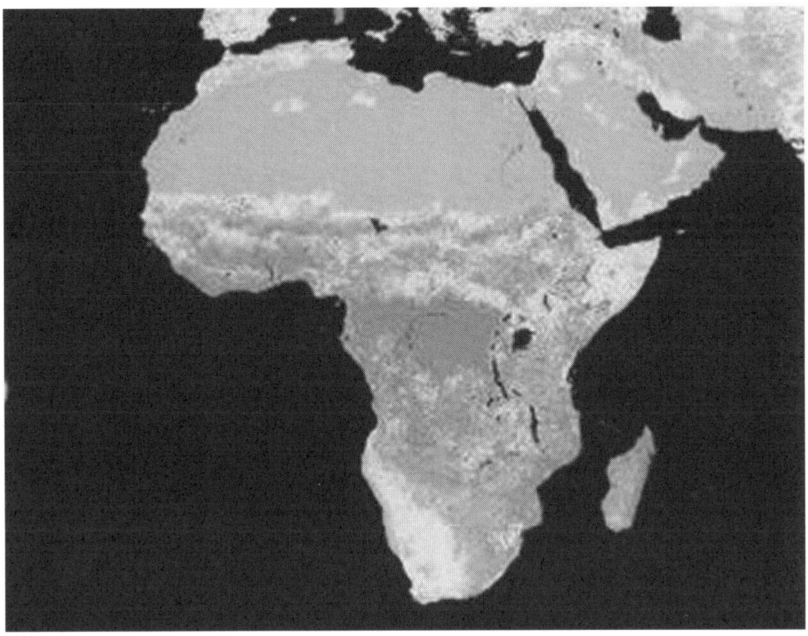

FIGURE 8.1
Protecting Africa's forest and savanna resources is a growing priority of development agencies like the African Development Bank. In satellite image, only the dark grey area just under the bulge and on its southern coast is forest. The cloud-like lighter areas are savannas. The light gray area across the northern half of the continent is desert or sparsely vegetated (NASA Goddard Space Center Scientific Visualization Studio, with permission).

WEB SITE www.afdb.org/

KEY WORDS desertification, forestry, renewable energy, water quality

PURPOSE to promote economic and social development through loans, equity investments and technical assistance in regional member countries.

SUMMARY *AFDB* has developed an integrated environmental and social screening system to identify and assess impacts of its projects. In addition, its *Sustainable Development Resource Center* provides guidance to staff on environmental policies, links to relevant environmental information networks and training to member counties to protect and restore the environment. It also houses the African regional office of the *U.N. Convention to Combat Desertification*. Recently approved projects for sustainable development include *The Regional Environmental Information Management* project for the Congo Basin focusing on biodiversity, small-scale renewable energy financing and forestry resource management in Benin, including alternative fuels for firewood. ADB also offers a searchable catalogue of publications. Most technical reports are sold ($10 each), but a few are free, including "How to Assess Environmental Impacts on Tropical Islands and Coastal Areas" (training manual).

MAIN RESOURCES OFFERED R2, R6, R10, R12, R14

TECHNOLOGY FUNCTIONS T2, T3, T4, T5

SECTOR RELEVANCE government, nongovernmental organizations (NGOs)

CONTACT

Communication Unit
01 BP 1387 Abidjan 01
Côte d'Ivoire
Tel: 225-20 20 41 18
Fax: 225-20 20 40 06
E-mail: comuadb@afdb.org

South Africa Department of Environment and Tourism (DEAT)

WEB SITE www.environment.gov.za/

KEY WORDS biodiversity, coastal zone management, conservation, ecotourism, environmental indicators

PURPOSE to contribute to improvement of the quality of life of all South Africans by promoting the sustainable development, utilization and protection of natural and cultural resources.

SUMMARY *DEAT*, besides enforcing environmental laws, has placed emphasis on laying the groundwork for sustainable development through improved environmental reporting and establishment of transfrontier conservation areas (TFCAs). DEAT is developing a core set of national environmental indicators. These indicators, some national, some regional or local, and some sectoral in scope, cover a wide range of environmental media, including inland and coastal waters, biodiversity, land use, climate, waste management and emerging issues such as genetically modified organisms. In conjunction with the cities of Cape Town, Durban, Johannesburg and Pretoria, DEAT produced the first-ever metropolitan state-of-the-environment reports. In addition, South Africa is seeking to link environmental and economic reporting through development of various aspects of natural resource accounts. To make

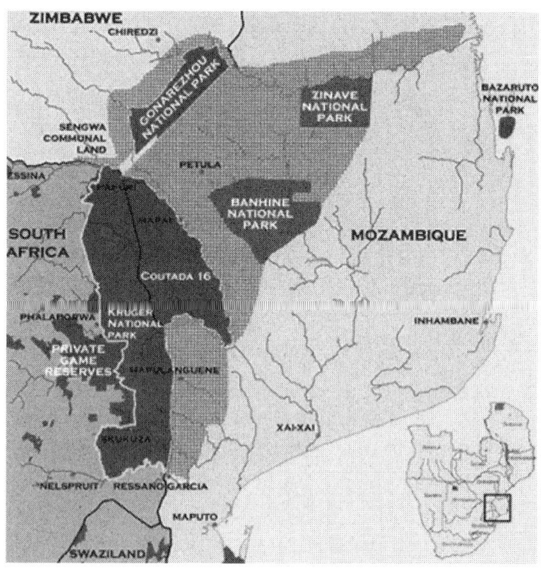

FIGURE 8.2
The new Great Limpopo Transfrontier Park, stretching over 35,000 square kilometers into South Africa, Zimbabwe and Mozambique, is to be managed to optimize benefits for biodiversity conservation and economic development of local communities.

better use of GIS data, the Internet and computer networks, DEAT has formed an in-house GIS forum to coordinate the different GIS activities in the department and to identify needs regarding access to relevant (environmental) geo-referenced data.

MAIN RESOURCES OFFERED R2, R10, R11, R12, R14, R15

TECHNOLOGY FUNCTIONS T1, T2, T3, T11, T12

SECTOR RELEVANCE business, NGOs, state and local government

CONTACT

Department of Environmental Affairs and Tourism
J.P. Louw, Communications Director
Tel: 012 310-3594
Fax: 012 322-2476
Mobile: 082 468-3401
E-mail: louwjp@iafrica.com

Association of South East Asian Nations (ASEAN)

WEB SITE www.aseansec.org/

KEY WORDS biodiversity, coastal waters, forests, transboundary haze

PURPOSE to share prosperity through human development, technological competitiveness and social cohesiveness.

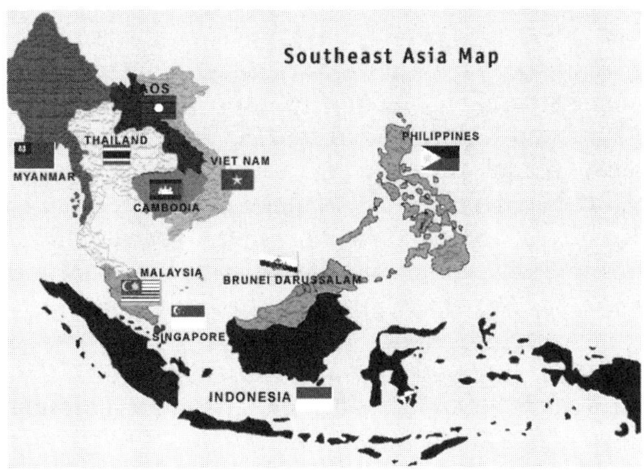

FIGURE 8.3
The ASEAN region has a population of about 500 million, a total area of 4.5 million square kilometers and a combined gross domestic product of US$737 billion (ASEAN).

SUMMARY *ASEAN* regional initiatives on the environment set out specific and measurable actions for each of the 15 objectives of the strategic plan of action on the environment. The areas of cooperation include: land and forest fires, transboundary haze, nature conservation and biodiversity, coastal and marine environment and global environmental issues. The ASEAN Regional Strategy for Haze Prevention and Mitigation, assisted by the Asian Development Bank (ADB) resulted in part in a transboundary haze pollution clearinghouse at www.haze-online.or.id. The clearinghouse facilitates coordination and monitoring of national, regional and international initiatives for preventing and mitigating forest fires and transboundary haze pollution. The ASEAN Haze Technical Task Force (HTTF) is addressing regional fire- and haze monitoring and prevention, and mitigation activities to deal with recurrences of regional haze, which occurred with widespread fire outbreaks in 1997.

MAIN RESOURCES OFFERED R1, R10, R12, R15

TECHNOLOGY FUNCTIONS T1, T2, T4

SECTOR RELEVANCE development banks, government

CONTACT

Bureau for Programme Coordination & External Relations
dhannan@aseansec.org

Asian Development Bank (ADB)

WEB SITE www.adb.org

KEY WORDS air quality, climate change, forests, greenhouse gases, transboundary haze, water quality

PURPOSE to promote sustainable development and environmental protection is a key strategic element of the bank's mission.

SUMMARY *ADB* offers loans and technical assistance to developing member countries (DMCs) for environment projects. ADB also offers technical assistance on transboundary environmental issues such as climate change, acid rain in northeast Asia and the impact of atmospheric haze brought about by forest fires in the countries of the Association of Southeast Asian Nations (see ASEAN entry). Its loan portfolio includes projects in the People's Republic of China to improve water quality and flood control rehabilitation for the Suzhou Creek, and in three cities of Shanxi Province to improve air quality. In Sri Lanka, a bank project is addressing coastal erosion and resource degradation. A project in Viet Nam addresses overall environmental conditions in Ho Chi Minh City. On climate change, ADB, the Global Environment Facility (GEF) through the United Nations Development Programme, and the Norwegian Government funded an Asia Least-Cost Greenhouse Gas Abatement Strategy. The study identified abatement measures that could reduce greenhouse gas emissions from the 11 countries by approximately one third of their projected "business as usual" emissions in 2020. Currently, the 11 participating countries, which have about half the population of the world, account for 12% of all global carbon dioxide-equivalent emissions.

MAIN RESOURCES OFFERED R2, R4, R6, R12, R14

TECHNOLOGY FUNCTIONS T1, T2, T3, T4, T6, T12

SECTOR RELEVANCE developing country governments

CONTACT

Asian Development Bank
P.O. Box 789
0980 Manila, Philippines
Tel: 632-632-4444
Fax: 632- 636-2444

(Japan) The National Institute for Resources and Environment (NIRE)

WEB SITE www.aist.go.jp/NIRE/index-e.htm

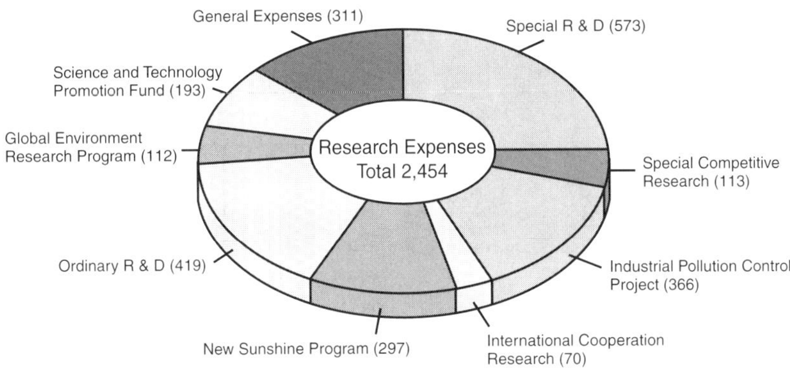

FIGURE 8.4
NIRE focuses on a broad range of environmental issues for sustainable development under the rubric of "ecotechnology."

KEY WORDS biodegradation, carbon sequestration, catalyst, combustion, dioxin, energy storage, enzyme, heating and cooling, photocatalysis

PURPOSE to pursue R & D to ensure sustainable development.

SUMMARY *NIRE* is concerned with a wide range of research fields such as environmental protection, natural resources development, new forms of energy, and industrial safety. It has adopted the concept of "ecotechnology," which combines "technology" with "renewable" and "ecologically" sound resources development. Although some information is posted on the Institute's Web site in English, NIRE generally requires formal arrangements for technology transfer, technical assistance and training. NIRE also offers group and individual training courses to developing countries under the auspices of the Japan International Cooperation Agency (JICA). Research programs currently cover technologies for maximum energy and resources utilization, environmental measuring and monitoring technologies, pollution control and treatment, and evaluation of environmental effects and ecotechnology. NIRE publishes a quarterly *Journal* and an *Annual Report* as well as special reports on selected topics.

MAIN RESOURCES OFFERED R1, R2, R12, R14, R15

TECHNOLOGY FUNCTIONS T1, T2, T3, T4, T6, T8, T19, T12

SECTOR RELEVANCE chemicals, developing countries, energy, manufacturing, state and local government

CONTACT

Technical Advice Center
Tel: +81-298-61-8105

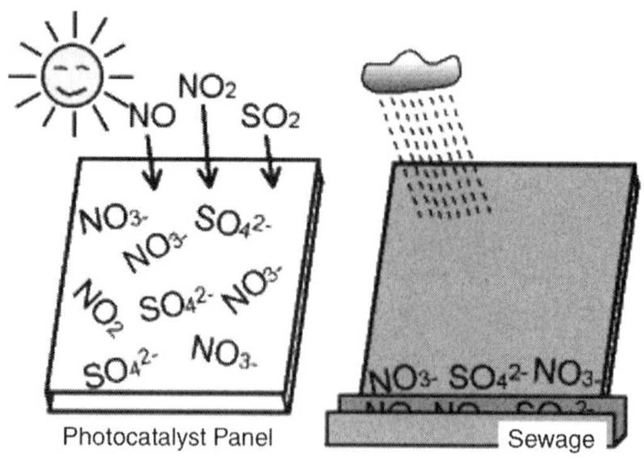

FIGURE 8.5
Photocatalysis. (K. Takeuchi, Air-purifying sheets for cleaner streets, *Look Japan*.)

Sample Technology: Atmospheric Purification by Photocatalysis (NIRE)

ENVIRONMENTAL CHALLENGE Atmospheric constituents such as chlorofluorocarbons (CFCs) and CFC substitutes, greenhouse gases and nitrogenous and sulfurous compounds undergo photochemical reactions either directly or indirectly in the presence of sunlight. In a polluted area, these can build up in the atmosphere rather than eventually being removed. NIRE is especially concerned with heterogeneous processes in which particulate matter and aerosols play a role.

PROJECT RESULTS Photocatalytic materials based on titanium dioxide (TiO_2) exposed to sunlight can remove nitrogen and sulfur oxides from ambient air to the levels of environmental standards (about 0.05 ppm), according to NIRE. As these photocatalysts are activated by sunlight and regenerated by rainfall, they purify ambient air naturally without additional energy use, according to test results. Some local governments have begun considering their use.

(Japan) National Institute for Environmental Studies (NIES)

WEB SITE www.nies.go.jp

KEY WORDS air quality, biodiversity, climate change, endocrine disruptors and dioxins, ozone layer, watershed management

PURPOSE to carry out interdisciplinary research in a variety of fields to respond to public awareness of the environment, public expectations and the new environmental management needs of the government.

SUMMARY Formerly the main research branch of the Environment Agency (EA) of the government of Japan, *NIES* became an independent institute in April 2001. It is pursuing six priority areas of research, such as ecosystem monitoring and the impact of nitrogen oxide and particulate matter. It offers databases and research reports, mostly in Japanese. The institute participates in international research and provides information to other countries, including developing countries. It also provides experts to the international technical cooperation programs conducted through the Japan International Cooperation Agency (JICA). A new initiative tied to implementation of the Kyoto treaty on global greenhouse gases is the AsiaFlux regional network to coordinate data collection. To implement the treaty, individual countries are required to estimate their greenhouse gas budgets and how much carbon is sequestered by forests and other means. The AsiaFlux network, launched in March 2000, is designed to contribute toward improving data collection techniques, establishing a database and promoting exchange of information. Scientists and modelers studying carbon dioxide, water vapor and energy fluxes are invited to join AsiaFlux, which has its own Web page linked to the Institute's site.

MAIN RESOURCES OFFERED R2, R6

TECHNOLOGY FUNCTIONS T1, T2, T3

SECTOR RELEVANCE Various

CONTACTS

NIES
16-2 Onogawa, Tsukuba-Shi, Ibaraki
305-8506
Japan
E-mail: www@nies.go.jp

For reports:
16-2, Onogawa, Tsukuba, Ibaraki
305-0053,
Japan
Tel: +81-298-50-2308
Fax: +81-298-51-2854
E-mail: kokusai@nies.go.jp

Environment Canada (EC)

WEB SITE www.ec.gc.ca

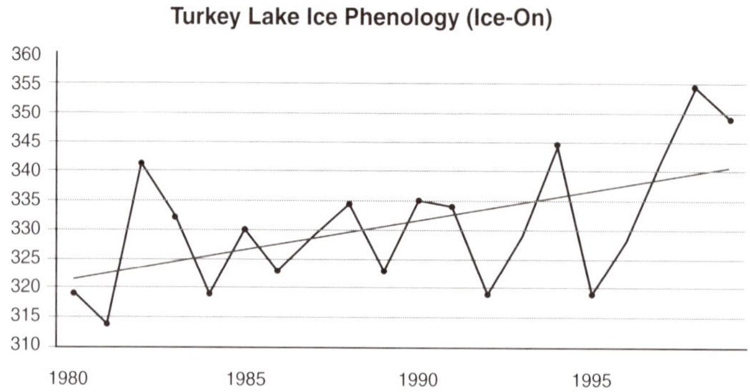

FIGURE 8.6
To detect climate-change effects, scientists, with the help of citizen observers, have tracked the degree of ice cover on rivers and lakes. Turkey Lake (about 50 kilometers north of Sault Ste. Marie, Ontario) shows a definite trend toward the lake's completely freezing over (winter ice-on) at later dates in the year.

KEY WORDS air quality, clean fuels, climate change, decontamination, endangered species, lakes and rivers, life-cycle analysis, materials substitution, manufacturing technology, mercury, product development, remediation, site assessment, solid waste, spill prevention control, wildlife refuge.

PROGRAMS

- EcoAction Community Funding Program
- Emergencies Engineering Technologies Office (EETO)
- Environmental Technology Advancement Directorate (ETAD)
- Environmental Technology Center (ETC)
- Ontario Centre for Environmental Technology Advancement (OCETA)
- Program of Energy Research and Development (PERD)
- Technology and Industry Branch (TIB)
- Technology Partnerships Canada (TPC)
- Technology Transfer Branch (TTB)
- Wastewater Technology Center (WTC)

PURPOSE to make sustainable development a reality in Canada by helping Canadians live and prosper in an environment that needs to be respected, protected and conserved.

SUMMARY *EC* embraces a wide range of environmental and natural resource functions, from meteorology to wildlife. In addition to compliance activities, EC programs encompass pollution prevention, global environmental issues and technology transfer. In some cases, the government charges for providing expertise to industry to develop products. EC offers information on hundreds of technologies that are at various stages of development or are commercialized. In some cases, the technology descriptions are accompanied by case studies or client testimonials. Resources vary from databases and clearinghouses to funding opportunities.

EcoAction Community Funding Program provides financial support for community projects that show positive effects on the environment.

EETO demonstrates and adapts innovative methods for on-site mitigation of water or soil contamination caused by pollution emergencies or insecure hazardous-waste sites. In conjunction with research partners, it evaluates response equipment and conducts product testing with products such as absorbent materials.
ETC develops and transfers new pollution measurement, prevention, control and remediation technology to foster clean air and mitigation of oil and hazardous-materials releases.
ETAD's four focus areas are 1) measuring air pollutants in ambient air and from mobile and stationary sources, 2) analyzing organic and inorganic compounds, 3) assessing and cleanup of contaminated sites and 4) preventing and responding to pollution emergencies.
OCETA offers links to profiles of cutting-edge Canadian environmental technologies (listed by environmental media) and companies providing environmental products or services in Ontario. (See OCETA Technology Profiles.)
PERD houses research, development and field testing in non-nuclear energy areas such as cleaner transportation fuels and systems.
TIB transfers technologies and expertise to the international community and conducts technology assessments through *Technology Partner Canada* initiatives.
TTB protects and commercializes EC's intellectual property and conducts outreach activities.
WTC develops and assesses new wastewater technologies and advancements.
TPC, a $300 million per year investment fund, focuses on partnering with industry to develop technologies that also create jobs.
CPPIC is a searchable database providing more than 1,200 P2 references ranging from fact sheets to case studies.

MAIN RESOURCES OFFERED R1, R3, R4, R5, R6, R10, R12, R13, R15

TECHNOLOGY FUNCTIONS T2, T4, T8, T10, T12

SECTOR RELEVANCE academia, international governments and agencies, NGOs, private sector

CONTACTS

General Information
Environment Canada:
Inquiry Centre
351 St. Joseph Boulevard
Hull, Quebec
K1A 0H3
Tel: 819-997-2800 or 1-800-668-6767
Fax: 819-953-2225
E-mail: enviroinfo@ec.gc.ca

Environmental Technology Center
3439 River Road South,
Gloucester, ON K1A 0H3
Tel: 613-991-9550
Fax: 613-998-1365

For information on funding programs:
www.ec.gc.ca/fund_e.html

Sample Resource: OCETA Technology Profile Samples

- Rapid Vegetation Establishment System (commercialized), Soil Enrichment Systems Inc. Proprietary system for the rapid establishment of vegetative environments in stressed soil systems.
- Rubber Manhole Head and Riser Assembly (commercialized, Enviromax Inc.). Manhole head and riser assembly made entirely of 100% recycled rubber, which improves the lifespan of manhole structures and the surrounding asphalt.
- Eco Logic Process (commercialized, Eco Logic International Inc.). Process to convert high-concentration organic contaminants, in solid or liquid form, into reusable products using a gas-phase chemical reduction reaction.
- Metal Separation by Liquid Ion Exchange (pilot stage, MET-TECH Systems Ltd.) Highly efficient liquid–ion exchange process for the recovery, separation and concentration of a wide range of heavy metals.

Case Study: Life Cycle Analysis of a Business Telephone

This case study summarizes a 2-year joint project between Nortel Networks and Environment Canada to analyze the material, energy and environmental aspects of a common consumer product — the telephone. The particular telephone chosen was the Norstar M3710, which at the time was produced by Nortel Networks. The goal of the project was to better understand the potential economic and environmental benefits of emerging environmental management concepts such as product life-cycle assessment and design for environment. Companies and, increasingly, national governments are using and promoting these tools to enhance the environmental performance of products. Because these tools also identify opportunities for improvements in energy and resource efficiency, they provide companies with opportunities to reduce costs and stimulate innovation.

Eco-Action Grants

Grants for projects that have measurable positive impacts on the environment go to community organizations such as environmental groups, aboriginal groups and councils, service clubs, associations, and youth and senior organizations. Projects require 50% matching funds or in-kind support from other sponsors. Priority for funding is given to projects that will achieve results in the following areas: clean air and climate change, clean water, and nature. Submission deadlines to the funding program are February 1st and October 1st annually. Sample projects include replanting shorelines, cleaning streams, conserving water and recycling. See Web site for application requirements and success stories.

European Bank for Reconstruction and Development (EBRD)

WEB SITE www.ebrd.com

KEY WORDS emissions, environmental assessment, sanitation, wastewater

PURPOSE to foster the transition toward open market-oriented economies and to promote private and entrepreneurial initiatives in the countries of Central and Eastern Europe and the Commonwealth of Independent States (CIS). A high priority is to promote environmentally sound and sustainable development through the full range of its activities.

SUMMARY The *EBRD* seeks to help the 27 countries it works with to implement structural and sectoral economic reforms, promoting competition, privatization and entrepreneurship. To implement environmental policy, the bank subjects investment and technical cooperation activities to environmental appraisal and seeks ways to integrate environmental benefits or improvements in projects. Since its founding in 1991, the bank has provided about 1 billion euro for municipal and environmental infrastructure services. Environmental components of projects have included emission-reduction technologies, wastewater treatment, and expenditures related to environmental action plans. The bank has also supported environmental rehabilitation of oil fields in a desert and scrubland near Baku, Azerbaijan, using low-toxicity water-based drilling muds and blowout-prevention techniques. The bank's *Project Preparation Committee (PPC)*, an integral part of the *Environmental Action Programme (EAP)* for Central and Eastern Europe, works with private co-financers and other international financing organizations in developing projects. Resources available on its Web site include: sub-sectoral environmental guidelines, currently covering 83 business activities in 17 industry sectors; procurement opportunity notices; *Factsheets* on environmental activities; and project case studies. The biannual *Environments in Transition* publication (in English and Russian) includes brief descriptions of the environmental aspects of new projects.

MAIN RESOURCES OFFERED R1, R2, R5, R12, R15

TECHNOLOGY FUNCTIONS T1, T2, T3, T4, T5, T10, T11, T12

SECTOR RELEVANCE agriculture, gas, governments, industry, mining municipal services, NGOs, petroleum

CONTACTS

European Bank for Reconstruction and Development
One Exchange Square
London EC2A 2JN, U.K.

General enquiries:
Tel: +44 20 7338 6372
Fax: +44 20 7338 6690
E-mail: harrisob@ebrd.com

Requests for publications:
Tel: +44 20 7338 7553
Fax: +44 20 7338 6102

E-mail: Pubsdesk@ebrd.com

Project enquiries/proposals:
Tel: +44 20 7338 6282
Fax: +44 20 7338 6102
E-mail: projectenquiries@ebrd.com

European Commission–Environment Directorate

WEB SITE europa.eu.int/comm/dgs/environment/index_en.htm

KEY WORDS biodiversity, climate change, pvc, waste

PURPOSE to maintain and improve the quality of life through a high level of protection of natural resources, effective risk assessment and management and timely implementation of community legislation.

SUMMARY The Sixth Environment Action Programme of the European Community 2001–2010 identifies four priority areas: climate change, nature and biodiversity, environment and health, and natural resources and waste. To achieve improvements in these areas, the new programme sets out five key objectives: 1) ensure the implementation of existing environmental legislation, 2) integrate environmental concerns into all relevant policy areas, 3) work closely with business and consumers to identify solutions, 4) ensure better and more accessible information for citizens and 5) develop a more environmentally conscious attitude toward land use. The commission itself does not offer technology resources but has links to the European Environment Agency and other resources.

Online publications available include: *Analysis of the EU Eco-Industries*, which provides estimates on the market for goods and services produced by the eco-industries of the 15 EU nations and the Candidate Countries; *Chemical Recycling of Plastics Waste* (PVC and other resins), describing existing technologies, both industrial plants and pilot projects; and *Options to Reduce Methane Emissions*, which gives options for the reduction of emissions from agriculture, waste, coal mining and the oil and gas industry. The publications page also contains PDF versions of the quarterly *Environment for Europeans*, covering current environmental news in the Union.

MAIN RESOURCES OFFERED R1, R2, R15

TECHNOLOGY FUNCTIONS Various

SECTOR RELEVANCE government, petroleum

CONTACT

Environment Directorate-General
European Commission
Rue de la Loi 200
B-1049 Brussels
E-mail: envinfo@cec.eu.int

European Environment Agency (EEA)

WEB SITE www.eea.eu.int/

KEY WORDS acidification, air quality, biodiversity, chemicals, climate change, noise, ozone waste

PURPOSE to support sustainable development and help achieve significant and measurable improvement in Europe's environment by providing timely, targeted, relevant and reliable information to policymaking agents and the public.

SUMMARY *EEA* acts as a clearinghouse for environmental data, indicators and information to the European nations. The Web site breaks down offerings by environmental issues, media, sectors, geographic regions and environmental solutions with which businesses, industrial leaders, environmental professionals, local authorities and others across Europe are experimenting. Some information and data resources offered include: *Environment Information and Observation Network (EIONET)*, serving 18 National Focal Points (NFP) responsible for national coordination of activities related to the EEA Work Programme (NFPs offer publicly available environmental information in their respective countries); *European Community Clearinghouse Mechanism for Biodiversity Service; European Topic Centre on Waste and Material Flows NATLAN* (NATure/LANd cover); *EnviroWindows,* covering environmental management concepts and sustainable development for NGOs, research groups, businesses and private enterprises; *SERIS* (State of the Environment Reports Information System) for individual European countries and *Computer Intelligent Processing Technologies (CIPTs)* to help environmental analysts select IT tools.

MAIN RESOURCES OFFERED R1, R2, R15

TECHNOLOGY FUNCTIONS T1, T2, T3, T5, T6, T8

SECTOR RELEVANCE academia, governments, industry, NGOs

CONTACT

EEA, Information Centre, Kongens Nytorv 6, DK-1050
Copenhagen K
Denmark
Tel: +45 33 36 71 00/68/65 (the two last extensions are direct lines to the IC)
Fax: +45 33 36 71 99
E-mail (preferred): information.centre@eea.eu.int

Global Environment Facility (GEF)

WEB SITE www.gefweb.org

KEY WORDS biodiversity, climate change, coastal waters, desertification, ozone layer

PURPOSE to forge international cooperation and finance actions to address critical threats to the global environment.

SUMMARY *GEF*, which counts 171 donor and recipient countries as members, addresses global environmental problems in four areas: biodiversity loss, climate change, degradation of international waters and ozone depletion. Desertification and deforestation projects, as they relate to these critical threats, are also included in its portfolio. GEF-funded projects are conducted through implementing agencies such as the World Bank, United Nations Development Program, United Nations Environmental Program and regional development banks. Project partners include NGOs, developing-country civil societies, and private enterprises. During its first decade since launching in 1991, GEF has allocated $4.2 billion in grants, which have been supplemented by co-financing. More than 800 projects have been launched in 158 developing countries. Aside from country-specific projects, it has fostered transboundary land and multinational coastal projects. GEF projects aim to establish models for replication, develop tools such as biodiversity and ecosystem indicators and incentives for private investments in biodiversity conservation and renewable energy. GEF free information resources include project factsheets, project performance reports, *GEF Lessons Notes* and the *Annual Report*. Some are available on its Web site. GEF-financed projects have aided the Russian Federation and Eastern European nations to reduce consumption of ozone-depleting substances by 90% and by switching to ozone-friendly technologies. GEF also has launched nearly $1 billion in energy efficiency and renewable energy projects in 45 countries, more than $1 billion for 345 biodiversity projects and over $500 million for water resources projects. (See sample projects that follow this entry for examples).

FIGURE 8.7
The Global Environment Facility works within developing countries and emerging independent countries to enhance their capacity to address global environment problems in the context of their national development plans (Global Environment Facility, with permission.)

MAIN RESOURCES OFFERED R1, R3, R6, R10, R12, R15

SECTOR RELEVANCE consultants, governments, NGOs, renewable energy industry

CONTACT

Hutton Archer
Senior External Relations Coordinator
GEF Secretariat
1818 H Street NW
Washington, DC 20433, USA

Sample Project: Conservation of the Dana and Azraq Protected Areas (GEF)

KEY WORDS biodiversity

ENVIRONMENTAL CHALLENGE With the expanding population, human encroachment on natural ecosystems threatens the diversity in genes, species and the ecosystems that support life. Unfortunately, the areas of fastest population growth contain some of the richest concentrations of biological species. All of the species that are near extinction exist in just 13% of the earth's land area. An example of this concentration is Jordan's Dana Nature Reserve, an arid and semiarid area that is home to 20% of Jordan's native floral species and rare animal species that are threatened by unsustainable resource use by local communities, as is the Azraq Oasis, where extraction of water for urban and agricultural use threatened to dry up the ecosystem that has provided a critical habitat to millions of waterbirds.

PROJECT OBJECTIVES to protect and conserve the biodiversity of the Dana wildlands and the Azraq Oasis and strengthen national conservation management and environmental education.

PROJECT RESULTS Since Jordan created special reserves for the Dana wildlands and Azraq Oasis, GEF projects have supported a pilot phase of rehabilitation through the Royal Society for the Conservation of Nature (RSCN), an NGO that undertook responsibility for both reserves and an ongoing phase by Jordan's government. Sample measures have included working out an agreement with a cement corporation and government mining agency to limit activities in the Dana reserve area, as well as restricting the grazing of the Bedouins' livestock there. In the Oasis project, a permanent water supply and water management plan is in place. Some medicinal plants that were disappearing have been making a comeback and RSCN has expanded its training and support to environmental groups throughout the region.

SECTOR RELEVANCE eco-tourism, NGOs

Contact See main GEF entry

Sample Project: Biodiversity Conservation in the Choco Region

KEY WORDS aquaculture, biodiversity, forests, medicinal plants

ENVIRONMENTAL CHALLENGE The Choco region of Colombia, encompassing 5600 square kilometers, stretches from the western slopes of the Andes Mountains to the Pacific Ocean. It possesses the greatest plant biodiversity on the planet. Twenty-five percent of the Choco's plant and bird species are not found elsewhere on earth. Local poverty, deforestation, mining, logging and possible industrial development threaten this rich biological heritage.

PROJECT OBJECTIVES to incorporate biodiversity concerns into Colombia's development plans, provide scientific information and identify alternative and sustainable use of biological resources

PROJECT RESULTS The Ministry of the Environment is fostering a development strategy incorporating biodiversity concerns and promoting sustainable uses of resources by local populations. Results include development of a comprehensive regional information system, training of local people as parataxonomists to create a biological information base, establishment of local community-led pilot projects on commercializing traditional medicines, aquaculture, fruits and other agroforestry products, and development of a biodiversity ownership policy to protect intellectual-property rights of local residents. (Source: project factsheet, GEF.)

SECTOR RELEVANCE forestry, logging, mining, pharmaceuticals

CONTACT See main GEF entry

Sample Project: Danube River Pollution Reduction Program (GEF)

KEY WORDS estuaries, eutrophication, toxic pollutants, watershed

ENVIRONMENTAL CHALLENGE High loads of nutrients, hazardous substances and other contamination and competition to draw water from the river contributed to degradation of the river and the Black Sea into which it flows. Along the 2800-kilometer river, its watershed comprises 13 countries with a combined population of 85 million people. Eutrophication of the Black Sea has led to reduction of the fish catch between 1986 and 1990 by nearly 90%. Underwater resources, such as the red algae *Phyloophora* shrank from 10,000 square kilometers to less than 1500. Other problems include introduction of non-indigenous species and oil spills.

PROJECT OBJECTIVES to develop prioritized pollution reduction projects, improve river basin protection policies and increase public awareness and participation.

PROJECT RESULTS GEF supported the founding of a multinational Environmental Program for the Danube River Basin (EPDRB) in 1991 and the Black Sea Environment Program in 1993. Among the project achievements have been the identification of "hot spots" for the Danube Pollution Reduction Program, which targets a 33% reduction in the phosphorous load and 22% in the nitrogen load carried by the river to the Black Sea. Follow-on measures under the GEF Strategic Partnership on the Danube/Black Sea Basin (involving UNDP, UNEP and the World Bank and 17 basin countries and the European Union) include further reductions in nutrients reaching the Black Sea, enhancing the ability of seabed plants to absorb nutrients and improving fisheries management. (Source: project fact-sheets, GEF Action in the ECE Region, Sept. 2001.)

SECTOR RELEVANCE fisheries, wastewater treatment

CONTACT see main GEF entry

Sample Project: Solar Off-grid Photovoltaic Systems (GEF)

KEY WORDS climate change, fossil fuel, renewable energy

ENVIRONMENTAL CHALLENGE About two thirds of the world's people have no access to electricity. Most of these people live in rural areas far from grid-connected power. They deplete forests that help mitigate emissions of carbon from the global use of fossil fuels.These characteristic rural energy sources have serious environmental effects, such as deforestation and toxic wastes associated with use of lead-acid batteries. Finding ways to bring distributed renewable energy to rural areas is a problem of adapting technologies and creating markets suitable to differing population needs and geographic regions. If the preponderance of the world's population continues in this pattern or adds to the fossil-fuel demand by migrating to urban centers, the carbon load on the atmosphere will inexorably rise.

FIGURE 8.8
GEF helps finance local private and nonprofit organizations to bring photovoltaic electricity to rural homes.

PROJECT OBJECTIVES to promote renewable energy by removing barriers, reducing implementation costs and reducing long-term technology costs. The goal is to promote development and use of technologies such as solar home systems (SHS) critical for addressing the climate change problem in the long term.

PROJECT RESULTS GEF has financed nearly $1 billion in energy efficiency and renewable energy projects at a total cost of more than $5 billion in 45 countries. Developing markets for rural SHS using photovoltaic panels has been a major thrust. GEF expects, for instance, that as a result of 21 projects to promote SHS, the number of systems installed (600,000) will have increased the installed base in developing countries globally by more than 100%. GEF has employed two basic private-sector-oriented models for delivery of SHS: dealer sales and energy service companies. In the former, a dealer purchases systems or components from manufacturers and sells them directly to households, usually as installed systems. In the service-company model, the company retains ownership for the SHS and charges a monthly fee to the household. Various forms of credits to homeowners or financial incentives to the SHS dealers have been employed to grow the market initially. These efforts have led to some indirect influences and impacts in some countries. In Sri Lanka, for instance, an energy services delivery project contributed to the entry of Shell International Renewables in the countries PV market. In Mexico, a renewable energy project and pre-project implementation studies and dialogues contributed to a decision by the Mexican government to shift more focus to rural renewable energy rather than further extensions of grid power. (Source: *Promoting Energy Efficiency and Renewable Energy*, GEF, June 2000.)

SECTOR RELEVANCE residential construction, solar energy

CONTACT See main GEF entry

Sample Project: Village-Based Management of Woody Savanna and Woodlots for Carbon Sequestration (GEF)

KEY WORDS climate change, greenhouse gases

ENVIRONMENTAL CHALLENGE Most households in Benin rely solely on firewood for cooking and other energy uses. Slash-and-burn agriculture practices are deforesting the country's woody savannas (which absorb carbon dioxide), at an estimated rate of 100,000 hectares yearly.

PROJECT OBJECTIVES to reduce CO_2 emissions from semiarid areas by better management of forests and village lands.

PROJECT RESULTS By helping the majority of local people to control site clearing, switch to energy-saving stoves and grow legume trees for animal feed, these measures are expected to prevent more than 5 million tons of carbon from release to the atmosphere. Local involvement includes educating farmers, herders, hunters and beekeepers on the crucial ecological role of trees; women on how to build energy-saving stoves and local groups on how to implement forest management plans. (Source: factsheet, GEF.)

FIGURE 8.9
A GEF/UNDP project in the West African Nation of Benin has developed alternatives to traditional slash-and-burn rural agriculture practices that have already reduced greenhouse gases (Bruno Locatelli, with permission).

SECTOR RELEVANCE agriculture, forestry, livestock

CONTACT

United Nations Development Programme
Mahenau Agha, Information Officer
One United Nations Plaza
New York, NY 10017 USA
Tel: 212-906-5004
Fax: 212-906-6998
E-mail: mahenau.agha@undp.org

Inter-American Development Bank (IADB)

WEB SITE www.iadb.org

KEY WORDS biodiversity, climate change, coastal zone management, conservation, environmental assessment, environmental management, forestry, natural resources, renewable energy, water quality

PURPOSE to help accelerate economic and social development in Latin America and the Caribbean and to further the goal of sustainable development by enhancing the environmental quality of bank operations.

FIGURE 8.10
Sustainable development is a major thrust of new IDB lending and grants (Inter-American Development Bank, with permission).

SUMMARY In carrying out its mission, *IADB* has mobilized financing for projects that represent a total investment of $263 billion, with lending currently at the rate of over $5 billion annually. In the past, the bank emphasized lending in agriculture and industry, the physical infrastructure sectors of energy and transportation and the social sector. Current lending priorities include the environment. In 2000, $531 million in loans for environmental and natural-resource development included programs for water and sanitation and urban development, natural-resources conservation, tourism, environmental management and natural disasters. IADB grants for addressing global environmental problems are expected to increase as much as $30 million annually under a new agreement with the Global Environment Facility (see IADB-GEF entry this chapter). Further, special projects in environmental management and energy efficiency offer additional resources. Through its *Sustainable Development Program* (SDP) (www.iadb.org/sds/index.htm), the bank posts information about technology-related resources in various sectors and countries, such as a training course in ecologically sustainable industrial development.

MAIN RESOURCES OFFERED R1, R2, R4, R6, R12, R14, R15

TECHNOLOGY FUNCTIONS T1, T2, T3, T4, T5, T6, T7, T8, T11, T12

SECTOR RELEVANCE agriculture, developing national governments, energy, industry, NGOs, private enterprises

CONTACT

External Relations
Inter-American Development Bank
1200 New York Avenue, N.W.
Washington, D.C. 205777
Tel: 202-623-1000
Fax: 202-623-3096

Sample Resource: Global Environment Facility (IADB)

FIGURE 8.11
The IDB plans to initiate biodiversity projects such as preservation of tropical forest trees that are home to great numbers of insect species (Roger Hamilton, IADB, with permission).

Key Words

biodiversity, climate change, greenhouse gases, ozone protection, water quality

ENVIRONMENTAL CHALLENGE Low-income countries, with many pressing needs, are often reluctant to request loans for projects that target such global concerns as biodiversity loss and global warming.

PROGRAM OBJECTIVES to address global environmental issues in developing countries of Latin America.

PROGRAM DETAILS Until recently, Latin American countries could not access GEF funding (see separate entry this chapter) through IADB. Under an agreement with GEF signed last year, they now can apply to IADB for GEF project grants in the areas of biodiversity protection, greenhouse gas reduction, ozone protection and international waters. These grants, ranging up to a maximum of $350,000, are expected to be a significant part of overall IADB financing for biodiversity, renewable energy and energy efficiency. The bank believes that total GEF grants could amount to $1.7–3.5 million annually.

SECTOR RELEVANCE developing country governments, NGOs

CONTACT See main IADB entry.

Sample Resource: Hemispheric Sustainable Energy and Transportation (HSET) Funds

FIGURE 8.12
IADB funds nonconventional renewable-energy applications such as solar panels and wind power (Photo Library, Inter-American Development Bank).

WEB SITE www.iadb.org/sds

KEY WORDS clean fuels, renewable energy

ENVIRONMENTAL CHALLENGE Many rural areas in Latin America cannot be served economically by grid-connected power, and air pollution has reached health-damaging proportions in major urban centers.

PROGRAM OBJECTIVES *HSET* promotes the establishment of country trust funds to finance non-conventional renewable energy applications (e.g., wind, photovoltaic, biomass) and clean and efficient urban transportation (e.g., clean fuels, efficient modes of travel and land use, efficient vehicles).

PROGRAM RESULTS The fund has begun receiving proposals. The U.S. Department of Energy (USDOE) has contributed $1.25 million to initiate the HSET funds. Active fundraising efforts are under way to solicit contributions from other donor governments. Proposals may be submitted to the nearest IADB country office, any of the operational departments, or the Sustainable Markets for Sustainable Energy (SMSE) Program, at IADB headquarters. Institutions eligible for funds include national governments and their political subdivisions, NGOs and private entities with the legal authority to borrow funds and receive technical assistance, and regional and subregional organizations whose member countries are beneficiaries of bank operations.

SECTOR RELEVANCE IADB member governments, NGOs, private entities

CONTACT

Sustainable Development Department
HSET Funds
1300 New York Avenue, N.W.
Stop W-500
Washington, D.C. 20577
E-mail: HSET@iadb.org
www.iadb.org/sds/hset/

Sample Resource: Sustainable Development Program for the Pantanal Wetland (Brazil) (IADB)

WEB SITE www.iadb.org/goto.pl?pantanal

KEY WORDS biodiversity, ecotourism, fish farming, nonpoint pollution, sustainable agriculture, watershed management, wetlands

ENVIRONMENTAL CHALLENGE Sediment and organic and chemical contamination from development encroachment on the Pantanal, the world's largest wetland (recently named a World Heritage Site and a World Biosphere Reserve).

PROGRAM OBJECTIVES to foster sustainable development in the Pantanal and Upper Paraguay River Basin.

PROGRAM FEATURES Initial 4-year funding for the loan to the Brazilian Ministry of the Environment of $85.25 million will implement water management and environmental protection systems in critical areas of the Pantanal watershed. Strategies include soil and pesticide management, urban sanitation, and training for women in indigenous areas. Phase 2 will address sewage problems in nine cities surrounding the Pantanal, sustainable agriculture practices, fish production and ecotourism.

ANTICIPATED PROJECT RESULTS Improved quality of life from providing basic water and sanitation, reducing of health risks by decreasing agro-toxic chemicals in the water and increasing income from agriculture producers.

SECTOR RELEVANCE agriculture, ecotourism, sanitation, soil conservation, water management

CONTACT

Ministry of the Environment
www.mma.gov.br/

Organization of American States (OAS)

WEB SITE www.oas.org/

KEY WORDS renewable energy, sustainable development

PURPOSE through sustainable development projects, to follow up on the mandates of the United Nations Conference on Environment and Development Agenda 21 and of the Bolivia Summit of the Americas on Sustainable Development.

PROGRAMS

- Unit for Sustainable Development and Environment (USDE)
- Inter-American Agency for Cooperation and Development (IACD)

SUMMARY The *OAS*, the world's oldest regional organization, has modest programs for technical cooperation and assistance. Recently, these have included projects in sustainable development. *USDE* is the principal technical arm relating to sustainable development. It seeks the best practices and qualified suppliers of one country, matches them with the specific demands of another country and then assists in obtaining grants and private–public-sector partnership arrangements to finance the ensuing project. Technical issues addressed include transboundary management of water resources, natural hazards mitigation, public participation in decision-making, climate change or sea-level rise, coastal-zone management, renewable energy planning and biodiversity. The unit currently manages a portfolio of $44 million, primarily from externally generated funds. Limited information on the technologies used in projects is available under the publications link. The *IACD*, as part of its best practices program, carries out technical assistance and education projects in member OAS states through a special multilateral fund that amounts to about $8 to $9 million annually. The bank is considering a large electrification and telecommunications initiative in rural areas utilizing renewable energy.

MAIN RESOURCES OFFERED R1, R2, R6, R12, R14

TECHNOLOGY FUNCTIONS T3, T8, T12

SECTOR RELEVANCE academia, environmental services, member governments, NGOs

CONTACT

Inter-American Agency for Cooperation and Development
1889 F Street N.W.
Washington, D.C. 20006
Tel: 202-458-3510
Fax: 202-458-3526
E-mail: agency@iacd.oas.org

Sample Resource: Inter-American Biodiversity Information Network (IABIN)

WEB SITE www.iabin-us.org

KEY WORDS biodiversity, forests

ENVIRONMENTAL CHALLENGE A very small number of countries account for a very high percentage of the world's biodiversity, with three of the top four, Brazil, Colombia and Mexico, being located in the Americas. The Western Amazonian Uplands is the world's richest area in terms of biodiversity. Significant areas of natural habitat remain in the Americas, but the threats to their biodiversity are globally significant. In just one decade, the 1980s, for example, more forest was lost in South America than anywhere else in the world.

PROJECT STATUS The project has been technically cleared for funding by the Global Environment Facility. When available, the funds will be transferred to the OAS, which will execute the project under the authority and direction of the IABIN Council, chaired by the U.S. Geological Survey, which currently hosts the Web site for the IABIN.

PROJECT OBJECTIVES An initiative of the Summit of the Americas, *IABIN* has as its objective developing an Internet-based platform to provide access to scientifically credible biodiversity information currently existing in individual institutions and agencies in the Americas. IABIN will also provide the tools necessary to draw knowledge from that wealth of resources, which, in turn, will support sound decision-making concerning the conservation of biodiversity.

ANTICIPATED PROJECT RESULTS A region-wide, coordinated effort, originating from the countries in the region, to allow better access to and use of their biodiversity information will improve the quality and perhaps expand the quantity of biodiversity conservation activities globally.

SECTOR RELEVANCE academia, development agencies, governments, NGOs

CONTACT

Richard Huber
Environment Department
Unit for Sustainable Development and Environment (USDE)
1889 F Street N.W.
Washington, D.C. 20006
Tel: 202-458-3236
Fax: 202-458-3560
E-mail: rhuber@oas.org

United Nations Development Programme (UNDP)

WEB SITE www.undp.org/seed/unso/

KEY WORDS deforestation, desertification, drinking water, wastewater management

PROGRAMS:
- Office to Combat Desertification and Drought (UNSO)
- Global Water Partnership
- Public Private Partnerships for the Urban Environment

PURPOSE to provide development advice, advocacy and grant support. UNDP is leading the United Nations effort in building national capacity for environmentally sustainable development. Environmental degradation affects the poor most severely, because they are especially vulnerable to disease and economic loss due to water contamination, land degradation and air pollution.

FIGURE 8.13
Assisting local populations to develop replanting efforts for lands devastated by deforestation is an initiative of UNDP designed to halt deforestation and create jobs.

SUMMARY The UNDP works with development agencies and international lending and grant authorities in environmental remediation and pollution-prevention activities, emphasizing capacity building in developing countries and partnerships involving business, government and NGOs. The *Office to Combat Desertification and Drought (UNSO)* provides technical and financial support to affected countries in implementing the United Nations Convention to Combat Desertification and Drought (CCD). The *Global Water Partnership* works to integrate freshwater, river and coastal sustainable water management by encouraging stakeholders at all levels to work together more effectively. The partnership is an international network open to all organizations involved in water-resources management, including governments of developing as well as developed countries, UN agencies, multilateral banks, professional associations, research organizations, the private sector and NGOs. UNDP supports the *International Programme for Technology Research in Irrigation and Drainage* (IPTRID) (see Sample Resource) and the International Centre for Capacity Building under the umbrella of the GWP. *Public Private Partnerships for the Urban Environment* addresses populations, poverty and environmental problems. The partnerships program supports the development of joint-venture companies and offers capacity building to local governments and private groups for crafting innovative and sustainable partnerships.

MAIN RESOURCES OFFERED R1, R2, R6, R10, R12, R14

TECHNOLOGY FUNCTIONS T1, T2, T3, T4, T5, T12

SECTOR RELEVANCE academia, developing nation governments, NGOs, private sector

CONTACT

Communication Office,
United Nations Development Programme
E-mail: aboutundp@undp.org

Sample Resource: International Programme for Technology Research in Irrigation and Drainage (IPTRID)

KEY WORDS sustainable development, water conservation

ENVIRONMENTAL CHALLENGE Water, a necessary resource for human development, is often found in poor quality and availability in developing countries, making socioeconomic development exceedingly difficult. Conservation policies to protect freshwater and marine and coastal systems are required to reverse the degradation trends of freshwater and aquatic resources.

PROJECT OBJECTIVES *IPTRID*'s objective is to promote the capacity of developing countries to develop indigenous and sustainable solutions and to improve approaches, technologies and management for the use of water for food security and improved livelihoods.

ANTICIPATED PROJECT RESULTS An increase in investments from loans and grants by multilateral and bilateral agencies to enhance research capacity, quality and application results in individual countries. IPTRID also hopes to establish new arrangements for its governance and management as an associate of the Global Water Programme (GWP).

SECTOR RELEVANCE academia, developing country governments, water management

CONTACT See main UNDP entry

United Nations Environment Programme (UNEP)

FIGURE 8.14
UNEP, through its participatory GEO process, produces this review of the state of the world's environment, published by Earthscan for the UNEP (*Global Environment Outlook 2000*, published by Earthscan for the United Nations Environment Programme, 1999).

WEB SITE www.unep.org

KEY WORDS biodiversity, ecosystem, nonpoint pollution abatement, ozone protection, sustainable development, watershed management

PROGRAMS:

- Global Programme of Action for the Protection of the Marine Environment from Land Based Activities (GPA)
- International Environment Technology Center (IETC)

PURPOSE to provide leadership and encourage partnerships in caring for the environment by inspiring, informing and enabling nations and peoples to improve their quality of life without compromising that of future generations.

SUMMARY *UNEP* encourages sustainable development through sound environmental practices everywhere. Its activities embrace issues from atmosphere and terrestrial ecosystems to early-warning and emergency response for environmental disasters and emergencies. Present priorities include: environmental emergency response, enhanced coordination of environmental conventions, fresh water, technology transfer and industry and focused support to Africa. UNEP has its own training and capacity-building activities run by Regional Centres for Training and Technology Transfer, and lists training courses offered by other organizations worldwide. The *Division of Technology, Industry and Economics (DTIE)* works as a catalyst for government, industry and business to develop and adopt environmentally sound policies, strategies, practices and technologies (see Sample Resource). The *Division of Environmental Policy Implementation* is developing a UNEP-wide database on best practices, such as soil- and water-conservation initiatives. The *GPA* has prepared ten regional assessments and nine regional programs of action on land-based activities. The GPA clearinghouse mechanism offers information on pollutant source categories by the relevant UN agencies and of regional prototype nodes in selected regions. The *IETC* in Osaka (www.unep.or.jp/) promotes the adoption and use of environmentally sound technologies and offers links to relevant technology databases globally.

MAIN RESOURCES OFFERED R1, R2, R4, R5, R12, R14

TECHNOLOGY FUNCTIONS T1, T2, T3, T6

SECTOR RELEVANCE government, industry, NGOs, research institutions

CONTACT

United Nations Environment Programme
United Nations Avenue, Gigiri
PO Box 30552
Nairobi, Kenya

Tel: (254-2) 621234
Fax: (254-2) 624489/90
E-mail: eisinfo@unep.org (UNEP Webmaster)

Sample Resource: Cleaner and Safer Production Programme (UNEP)

WEB SITE www.uneptie.org/pc/cp/home.htm

KEY WORDS food chain, industrial processes, solid waste

ENVIRONMENTAL CHALLENGE Industrial development will be truly sustainable only if it is built on firm ecological foundations.

PROJECT OBJECTIVES The *UNEP DTIE Production and Consumption Unit (P&C)* works with international agencies, industry associations, and institutes to promote global awareness and understanding of sustainable production and consumption in industrial sectors with high environmental and safety impacts. Activities include:

- Assisting environmentally sound technology transfer through information exchange
- Capacity building and training
- Developing sound environmental management procedures in industry
- Creating international expert networks
- Linking with technical and policy bodies in key industry sectors

PROJECT ACTIVITIES The *P&C Unit* improves access to successful applications of sustainable development concepts in different industry sectors. Information resources include technical guides, policy reports, training manuals and case studies, the *Industry and Environment Quarterly* and the *International Cleaner Production Information Clearinghouse.* Publications for sale (order forms on the Web site) include *Cleaner Production: A Guide to Sources of Information* and sector reports for dairy, meat and fish processing, and pulp and paper mills. Aside from publications, training and workshops, PCU supports national cleaner production centers. As part of UNEP's input to the U.N. World Summit for Sustainable Development, UNEP is producing a status report on sustainable production and consumption.

SECTOR RELEVANCE food processing, manufacturing, mining, petroleum, pulp and paper mills, tourism

CONTACT

Tour Mirabeau
39-43 quai André Citröen
75739 Paris Cedex 15
France
Tel: (33 1) 44 37 14 50
Fax: (33 1) 44 37 14 74
E-mail: spchandak@unep.fr

World Bank—Environment Department

WEB SITE www.worldbank.org/environment/

KEY WORDS carbon sequestration, ecosystem, environmental indicators, forests, sustainable development

PROGRAMS

- Alliance for Forest Conservation and Sustainable Use (AFCSU)
- Critical Ecosystem Partnership Fund (CEPF)
- Prototype Carbon Fund (PCR)
- Environmental Economics and Indicators (EEI)

PURPOSE by linking poverty alleviation and sustainable development, to ensure that economic growth does not come at the expense of the world's physical and ecological systems or the world's poor.

SUMMARY In many developing countries, the cost of environmental degradation has been estimated at 4–8% of gross domestic product. The bank estimates some 6 million people in these countries die annually from water-related diseases, air pollution and toxic chemical exposure. The bank's environmental strategy aims to improve the quality of life and growth in these countries and protect regional and global environmental assets. The World Bank's portfolio of projects under Environmentally and Socially Sustainable Development amounted to US$16 billion as of the end of the 2001 fiscal year. About 95 active stand-alone environmental projects made up about one third of the portfolio, the remaining two thirds going to country or multi-country-sector projects with environmental objectives and global environmental projects. The bank targets programs and projects to foster long-term environmental sustainability, strengthen environmental management capacity and improve environmental conditions.

Aside from working environmental elements into projects, the bank is supporting some specialized environmental initiatives. It has financed over 558 projects to help enterprises convert to ozone-friendly technologies. The AFCSU, with the World Wildlife Fund, seeks to protect 125 million acres of highly threatened forest area around the world by 2005. The CEPF, involving Conservation International, the Global Environment Facility and the MacArthur Foundation, seeks better protection of biodiversity in hotspots occupying only 1.4% of the planet's total surface area but containing some 60% of all terrestrial species. The PCF develops real-world experience on how carbon markets and trading could operate in developing countries and in countries with economies in transition.

The bank has no specific inventory of environmental technologies employed on its projects, but various bank publications offer information to developing countries, such as the *Pollution Prevention and Abatement Handbook* (1998). By initiating a search for environmental technology in the bank's document base, one can find working papers, project appraisals, evaluations and special studies that describe types of technology employed in various projects, with some indication of overall results achieved. Many of the bank's projects now focus on capacity building for environmental assessments and monitoring, pollution control and pollution prevention (see Sample Resource).

Finally, the bank's EEI offers environmental economic tools and external training on environmental economics. The EEI team develops methods and applications of greener national accounts in the context of technical assistance to client countries and publishes annually the *Little Green Data Book* (see link under Contacts). It also supports the preparation of the annual estimates of "genuine" saving (the true saving rate of a country accounting for depreciation of produced assets, depletion of natural resources and human capital formation). These data appear in *World Development Indicators*. For an annual roundup of the bank's environmental activity, see *Environment Matters*.

MAIN RESOURCES OFFERED R1, R6, R12, R14, R15

TECHNOLOGY FUNCTIONS T1, T2, T3, T4, T5, T6, T11, T12

SECTOR RELEVANCE developing country governments, NGOs, nonprofits

CONTACTS

For publications:
Environment Department Publications
The World Bank
1818 H Street, N.W. Room MC 5-126
Washington, D.C. 20433
Tel: 202-473-3641
Fax: 202-477-0565

For Little Green Data Book
Hard copy, $15, order at:
publications.worldbank.org/ecommerce/catalog/product?item_id=1037953

Sample Resource: Mediterranean Environmental Technical Assistance Program (METAP)

WEB SITE lnweb18.worldbank.org/mna/mena.nsf/f34b224d37365b3f852567ee0068bd93/46bcc40bfb210a3a85256949006decce?OpenDocument

KEY WORDS biodiversity, coastal zone management, solid waste, wastewater, water quality

ENVIRONMENTAL CHALLENGE In most of the countries of the Middle East and North Africa, poverty and environmental degradation go together, especially in rural areas. Urban pollution, particularly regarding hazardous and solid waste and water quality management, is common among the METAP beneficiary countries.*

PROJECT OBJECTIVES to support the desire of the Mediterranean region to achieve sustainable development by assisting in project preparation and strengthening capacity in selected regional environmental management activities.

PROJECT RESULTS Established in 1990, *METAP* currently is supported by the European Commission, European Investment Bank (EIB), United Nations Development Program (UNDP), the World Bank, Switzerland and Finland. Projects totaling about US$2 billion have focused on water, wastewater and municipal solid waste, biodiversity and coastal-zone management. See Table 8.1 for a list of recent projects. According to a recent evaluation, METAP's major achievements include: 1) establishing a strong regional network of environment and technical specialists; 2) capacity building, feasibility studies and project preparation; 3) preparing environmental projects, many with innovative features, chiefly for EIB financing; 4) expanding the stakeholder base to include an increasing number of NGOs and community-based entities; 5) establishing 40 regional programs for knowledge

* Currently, these include Albania, Algeria, Bosnia and Herzegovina, Cyprus, Croatia, Egypt, Jordan, Lebanon, Libya, Morocco, Slovenia, Syria, Tunisia, Turkey, West Bank and Gaza.

sharing and 6) establishing a network of national focal points to coordinate their respective country activities.

SECTOR RELEVANCE development banks, developing country governments, NGOs

CONTACT

J.B. Collier
Mediterranean Environmental Technical Assistance Program (METAP)
Tel: 202-473-8551
E-mail: Jcollier@worldbank.org

World Bank Institute (WBI)

WEB SITE www.worldbank.org/wbi

KEY WORDS air quality, climate change, sustainable development, transportation planning

PURPOSE to create learning opportunities for countries, bank staff and clients, and people committed to poverty reduction and sustainable development.

SUMMARY *WBI* is the learning arm of the World Bank that assists countries in acquiring knowledge essential to the success of lending programs for sustainable economic development.

WBI has nearly 500 training activities annually for decisionmakers and policymakers in 149 countries on a variety of development and environmental policy topics. Open and distance-learning technologies are employed as well as in-person training. Its *Sustainable Development* program focuses on resources and activities that link poverty, environment, natural-resources management and governance. WBI also fosters partnerships for sustainable change and development that include governments, the private sector, NGOs, legislators and academics. The *Clean Air Initiative,* for example, facilitates knowledge sharing and offers technical assistance for clean air action planning (see Sample Resource).

MAIN RESOURCES OFFERED R1, R2, R4, R12, R14, R15

TECHNOLOGY FUNCTIONS T1, T5

SECTOR RELEVANCE developing country governments, NGOs, state and local governments

CONTACT

Communications Officer, Resources and Operations Unit,
World Bank Institute
1818 H Street, N.W.
Washington, D.C. 20433
Tel: 202-458-2498
Fax: 202-676-0858 (Office of the Vice President)
E-mail: wbi_infoline@worldbank.org

Sample Resource: Clean Air Initiative—Latin American Cities (WBI)

WEB SITE www.worldbank.org/wbi/cleanair/

TABLE 8.1

EIB Managed METAP III Investment Projects (Thousands of Euros)

EGYPT	
Cairo Sludge Disposal II (Completed)	63
Alexandria Wastewater Sludge & Effluent Reuse (Ongoing)	398
Alexandria Wastewater Sludge Extension (Ongoing)	49
GREECE	
Marina EIA (Identified)	250
Coastal Zone Assessment (Identified)	250
Olympic Village Sewerage (Identified)	500
JORDAN	
Detailed Design Study — Greater Amman (Completed)	636
Greater Amman EIA (Completed)	45
LEBANON	
Reuse of Sludge from Potable Water Treatment (Ongoing)	189
MOROCCO	
Settat Sanitation (Completed)	150
Settat II Sanitation (Ongoing)	150
Marrakech Sanitation (Ongoing)	500
Rabat Sanitation (Ongoing)	196
Tangier Sanitation (Ongoing)	200
Mohammedia Thermal Power Environmental Impact (Committed)	300
Nador (Identified)	300
TUNISIA	
Disposal of Municipal Solid Waste (Committed)	70
TURKEY	
Adana Water and Wastewater Tariffs (Completed)	26
Mersin Wastewater (Ongoing)	511
Diyarbakir Sewage II (Ongoing)	450
Marmara Region Wastewater (Ongoing)	345
Izmir Municipality (Committed)	360
Samsum Municipality (Committed)	395
Meric Maritsa River Basin UMMR (Identified)	250
Ismit Gulf Complementary Sewerage (Identified)	395
WEST BANK & GAZA	
West Bank Water and Sanitation (Completed)	140
TOTAL Luxembourg Managed EIB Investment Projects	7118

KEY WORDS air quality, clean fuels

ENVIRONMENTAL CHALLENGE Urban air pollutants that exceed the World Health Organization's (WHO) recommended ambient air quality standards cause about 200,000 deaths per year in Latin American cities, as well as the loss of billions of dollars annually due to health care costs, reduced quality of life and reduced economic investment, according to WHO.

PROJECT OBJECTIVES to improve air quality in Latin American cities

PROJECT RESULTS Since the program began in 1998, some cities show evidence of a downward trend in the highest concentrations of particulate matter and ozone. Pollution sources and progress vary from city to city.* The main thrust of the effort has been to increase the capacity of the cities to address air quality problems. Technical assistance to air quality action plans, in conjunction with private sector and development bank partic-

* The six largest cities in the region that are formal partners in the program include Buenos Aires, Lima, Mexico City, Rio de Janeiro, Santiago de Chile and São Paulo, but other cities have participated in workshops and courses.

ipation, has indirectly facilitated lending for environmental management and urban transport in the cities. The knowledge network established is a primary source of information and technical guidance on air quality management for other cities in the region. Internationally, the Latin American program has motivated the creation of three parallel regional networks in Africa, Eastern Europe and Central Asia and Southeast Asia. A formal external evaluation of the program is being conducted.

SECTOR RELEVANCE automobile manufacturing, developing-country governments, state and local governments

CONTACT

Clean Air Initiative in Latin American Cities

1818 H Street, N.W.

Washington, D.C. 20433

Tel: 202-458-0859

Fax: 202-676-0977/8

email: Clean_Air@worldbank.org

9

Regional, State and Local Agencies

Environmental Council of the States (ECOS)

WEB SITE www.sso.org/ecos/

KEY WORDS air quality, pesticides, pollution prevention

PURPOSE to champion the states' role in environmental management and to foster cooperation, coordination and exchange of information on environmental management.

SUMMARY: *ECOS* is mainly a policy advocacy and coordinating mechanism for state environmental agency enforcement and for cooperation with the U.S. EPA. It does, however, encourage and publicize innovative state pollution-prevention programs and helped establish the Interstate Technology and Regulatory Cooperation (ITRC) program described in this chapter. *Ecostates* is its primary publication (available on line and in print). Examples of innovative technologies ECOS has reported include a technology developed and marketed by Sensors, Inc, Saline, MI, to measure automotive tail pipe exhaust flow and mass emissions (the test design enables a decentralized testing (for report on its use, see NY Enhanced Vehicle Inspection in Contacts) and the Nebraska Pesticide Container Recycling Project, a voluntary program that has been recycling more than 100,000 containers annually. The council and EPA are also cooperating in creating the *National Environmental Information Exchange Network,* a voluntary, standards-based, Web-based network designed to improve the quality of environmental data and facilitate public access to it. EPA solicits applications for grants to states, the District of Columbia, trust territories and federally recognized Indian tribes that can be used to build capabilities for participating in the network (see Contacts).

MAIN RESOURCES OFFERED R1, R2, R4, R15

TECHNOLOGY FUNCTIONS T1, T2, T7, T11, T12

SECTOR RELEVANCE environmental equipment, state and local government, tribal government

CONTACTS

Environmental Council of the States
444 N Capitol St, NW
Suite 445
Washington, DC 20001
Tel: 202-624-3660
Fax: 202-624-3666
E-mail ecos@sso.org

Enhanced Vehicle Inspection/Maintenance Program

Department of Environmental Conservation

Web Site: www.dec.state.ny.us

For a copy of the annual report submitted to the U.S. EPA on the performance of the program, click on Air Resources under Environmental Protection, then on Enhanced Inspection and Maintenance under Key Programs and on Annual Report under Program Services.

EPA Information Network Grant Program:

http://www.epa.gov/neengprg/

EPA Regional and Tribal Programs

WEB SITE http://www.epa.gov/epahome/aboutepa.htm#regiontext

KEY WORDS coastal communities, coastal habitats, coastal waters, coastal zone management, landscaping, remediation, urban planning

PURPOSE Each of the ten EPA Regional Offices is responsible within its states for the execution of the agency's programs.

SUMMARY In addition to implementing national agency programs, some regions have special programs and projects with individual states or groups of states within their region or with states in adjoining regions. The URL above has links to all the regions, a map and a list of the states within each region. Examples of special programs, projects, or initiatives are described below. In addition, EPA has launched the National Environmental Information Exchange Grant Program to encourage state and other partners' data integration efforts.

- Region 3. This region sponsors Green Landscaping, a website for property owners interested in environmentally friendly gardening and landscaping. The average homeowner uses pesticides and fertilizers, often at rates 10-20 times that of farmers, and dumps pesticides rather than taking them to hazardous waste collection sites. In addition, EPA estimates tens of millions of gallons of gasoline are spilled each year while refueling garden power equipment. The site offers guidance on environmentally friendly practices such as composting yard waste, using a greater variety of plants, reducing the use of power equipment, and using plants to reduce heating and cooling needs
- *Regions 4 and 6. The Gulf of Mexico Community-Based Restoration Partnership (CBEP).* The CBEP is an agency-wide program. The two regions in conjunction with other federal agencies and five Gulf states make up the partnership committee, which receives input from the environmental, agricultural, tourism, fisheries, business and industry sectors of each state. The focus is on eight specific issues: habitat degradation, freshwater inflow, nutrient enrichment, marine debris, coastal and shoreline erosion, living aquatic resources, toxic substances and pesticides, and public health. The Gulf of Mexico Program Office is soliciting proposals for funding consideration (see Contacts).
- *Region 5. Antidotes to Sprawl.* The Region 5 site lists contacts for federal agency assistance on local land-use projects and decisions. Over time, many federal agencies and programs have actually encouraged sprawling development patterns. The programs listed, however, are offered as tools for "antidotes to

sprawl." The programs can be searched by specific issues such as transportation, flooding and habitat, and open space protection.

- *Region 8. Policy for Environmental Protection in Indian Country.* The principles of the Indian Policy state that Region 8 will, among other goals, work with tribal governments to protect human health and environment in Indian Country. Within EPA headquarters, the American Indian Environmental Office (AIEO) provides multimedia program development grants to tribes; negotiates agreements that identify tribal priorities for building environmental programs and direct EPA assistance; and develops tools to assist tribal environmental managers in setting environmental priorities.

MAIN RESOURCES OFFERED R2, R6, R10, R12, R14, R15

TECHNOLOGY FUNCTIONS T1, T3, T10, T11, T12

SECTOR RELEVANCE industry, nonprofit organizations, nurseries, state and local government, tourism, tribal government

CONTACTS COPY 167A

Region 3 (Delaware, District of Columbia, Maryland, Pennsylvania, Virginia, and West Virginia)

US EPA Region 3

1650 Arch Street (3CG00)

Philadelphia, PA 19103-2029

Tel: 215-814-5000 (Main Business Number) or 1-800-438-2474 (within region only).

Region 4 (Alabama, Florida, Georgia, Kentucky, Mississippi, North Carolina, South Carolina, Tennessee)

Gulf of Mexico CBEP

Larinda Tervelt

Gulf of Mexico Program Office

Bldg. 1103, Room 203

Stennis Space Center, MS 39529

Tel: 601-688-1033

Fax: 601-688-2709

www.epa.gov/gmpo/

Region 5 (Illinois, Indiana, Michigan, Minnesota, Ohio, Wisconsin and 35 Tribes)

U.S. EPA Region 5

77 W. Jackson Blvd.

Chicago, IL 60604

Tel: 312-353-2000

Or: 1-800-621-8431 (only within the region's states)

For Antidotes to Sprawl:

www.epa.gov/Region5/

Region 8 (Colorado, Montana, North Dakota, South Dakota, Utah, Wyoming and 27 Tribal Nations)

EPA Region 8 Office
999-18th St. Suite 300
Denver, CO 80202-2466
Tel: 303-312-6312 or 1-800-227-8917 (within region only)
www.epa.gov/Region8

National Environmental Exchange Grant Program
www.epa.gov/neengprg/

Interstate Technology and Regulatory Cooperation (ITRC)

WEB SITE www.itrcweb.org/common/default.asp

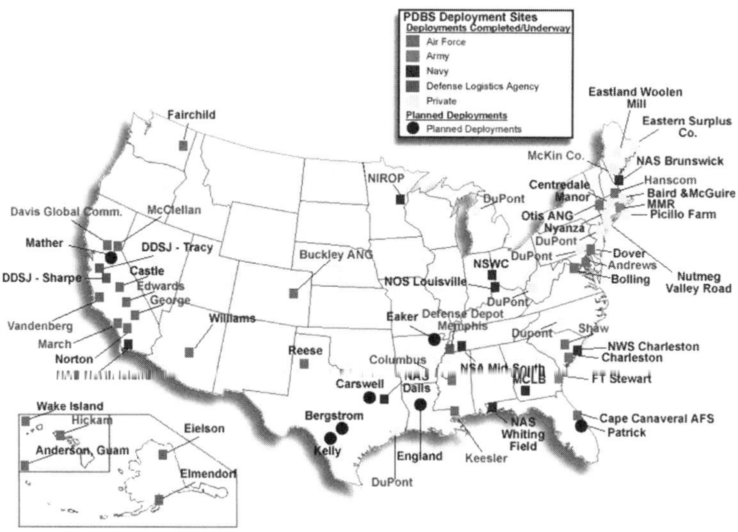

FIGURE 9.1
ITRC has assisted the military services and DOD to deploy the passive diffusion membrane sampler, a device for measuring groundwater contaminants. See description of the sampler in Chapter 5, Department of Defense under SERDP. Squares indicate deployments completed or under way; round dots indicate planned deployments.

KEY WORDS remediation, technology transfer

PURPOSE to facilitate the use of new technologies and help states maximize resources by breaking down barriers and reducing compliance costs.

SUMMARY *ITRC* is a state-led coalition working with industry, federal agencies and stakeholders to achieve regulatory acceptance of environmental technologies. More than 35 states work through ITRC. ITRC's site lists success stories in cutting approval time, slashing compliance costs, finding better solutions and paving the way for new technologies, which so far have focused on site remediation. Michael Duchesneau of Parsons

Engineering Services describes ITRC PRB documents as "very valuable and state-of-the-art guidance," in the design of permeable reactive barriers (PRB) at the Seneca Army Depot. ITRC also sponsored training on PRBs for both regulators and consultants working on the project. This shared experience helped create common ground and ultimately saved time for both parties in structuring an acceptable application. The Success Stories link provides a table of 84 projects involving site remediation technologies.

MAIN RESOURCES OFFERED R1, R3, R4, R12, R14, R15

TECHNOLOGY FUNCTIONS T1, T10

SECTOR RELEVANCE environmental equipment, state and local government

CONTACT

E-mail: itrc@wpi.org

National Governors Association (NGA)

WEB SITE www.nga.org

KEY WORDS brownfields, oil spill, remediation

Purpose to provide representation for states before the federal government, to provide networking seminars for state executive officials and to develop policy on reports on innovative state programs.

SUMMARY *NGA,* besides representation and policy development, promotes innovative state programs through its *Center for Best Practices.* The center promotes best practices in a number of areas, including the environment. Since 1994, for example, the center has worked with the U.S.EPA to better coordinate and promote state oil-spill-prevention programs. In 2001, NGA ran a special conference on redevelopment of brownfields, the approximately 450,000 former industrial and commercial sites across the United States that are contaminated, unused and often abandoned. It also has developed a series of reports on innovative site redevelopment programs and projects. Most center publications appear on the Web site http://www.nga.org/center/ in their entirety, including the newsletter *Front & Center.* Publications are available to the public, some at a nominal fee, but all are available at no cost to governors and their staffs.

MAIN RESOURCES OFFERED R1, R2, R4

TECHNOLOGY FUNCTIONS Various

SECTOR RELEVANCE state and local government

CONTACT

NGA Center for Best Practices
Hall of States
444 N. Capitol St.
Washington, D.C. 20001-1512
Tel: 202-624-5300

Local Government Organizations

WEB SITE See Contacts for Web sites

KEY WORDS environmental purchasing, nonpoint pollution, smart growth, urban planning, water quality, wetlands

SUMMARY To traditional services such as water sanitation and solid waste removal, city and county governments now face the whole array of environmental challenges, ranging from brownfield restoration to preserving biodiversity. Sometimes lacking the specialized expertise of national and state organizations, they have banded together to form their own mutual environmental assistance networks. The chief public-interest groups (PIGs) or associations representing the professional and elected communities of local government officials who deal with environmental issues as part of their responsibilities include the *International City/County Management Association (ICMA), National Association of Counties (NAC), National League of Cities (NLC), Public Technology, Inc. (PTI)* and *U.S. Conference of Mayors*. The PIGs all offer supportive information for implementing environmental assessments and adopting best practices. They often partner with the U.S. EPA, other federal agencies, or local government specialized professional associations. Examples include:

FIGURE 9.2
Local governments, old and new, were on the front line of environmental protection before the term was invented. (Top, Macon, GA; bottom, Rosells, IN, with permission.)

- *Local Government Environmental Assistance Network* administered by ICMA with other PIGs and supported by EPA (see Sample Resource).
- Urban Consortium Environmental Task Force (UCEnvTF) administered by PTI (see Sample Resource).
- Five-Star Restoration Challenge Grant Program, collaboration between NACo, the National Fish and Wildlife Foundation and the Wildlife Habitat Council, provides modest financial assistance to community-based wetland and riparian restoration projects, with major funding from the U.S. EPA and the National Marine Fisheries Service.
- *Smart Growth Network,* a coalition that includes NACo (see link from its Web site) and several private–public partners, which collaborates with U.S. EPA's Urban and Economic Development Division on outreach programs, research, publications and other projects on smart growth issues.
- NACo's Environmentally Preferable Purchasing Project and Water Program
- *Joint Center for Sustainable Communities,* a USCOM/NACo collaboration, to promote community leadership initiatives, public–private partnerships, technical assistance and training and policy and educational forums.

MAIN RESOURCES OFFERED R1, R2, R4, R6, R10, R14, R15

TECHNOLOGY FUNCTIONS T1, T3, T4, T5, T10, T11, T12

SECTOR RELEVANCE state and local government

CONTACTS

National League of Cities
www.nlc.org

International City and County Management Association
www.icma.org

National Association of Counties
www.naco.org

Public Technology, Inc.
www.pti.org

U.S. Conference of Mayors
usmayors.org/uscm/

Sample Resource: Local Government Environmental Assistance Network (LGEAN)

WEB SITE http://www.lgean.org

KEY WORDS solid wastes, urban planning, wastewater treatment, water quality

ENVIRONMENTAL CHALLENGE City governments often face state or federal mandates for environmental compliance and are the first level of government that citizens turn to for relief from pollutants affecting their lives. Yet budgetary constraints make it difficult for officials in small—and even large—jurisdictions to access impartial information, such as whether newer technologies have been proven in applications elsewhere.

PROJECT OBJECTIVES to provide a "first-stop shop" for environmental management, planning, funding and regulatory information keyed to the needs of local government elected and appointed officials and managers and their staffs.

PROJECT RESULTS Through the membership of its public interest group partners, the network is able to reach more than 100,000 local government officials and environmental professionals. LGEAN provides searches by keyword and three topic areas: local government operations, environmental media and environmental law. The network's Virtual City link allows the user to search an illustrative city for a range of environmental-impact areas, such as drinking water, wastewater treatment, landfills and emergency management. The site records more than 230,000 hits every month, and last year, LGEAN received more than 700 requests for information.

CONTACT

Local Government Environmental Assistance Network
Tel: 1-877-TO-LGEAN (865-4326)

Sample Resource: The Urban Consortium Environmental Task Force (UCEnvTF)

WEB SITE www.pti.org/task_forces/environment/index.html

KEY WORDS brownfields, monitoring, remediation, risk management

PROJECT OBJECTIVES To research and develop technology-based solutions that promote the best possible local government operations.

FIGURE 9.3
Cities have become proving grounds for many forms of environmental technology in partnership with organizations such as Public Technology, Inc.

PROJECT RESULTS Established in 1988, the task force works cooperatively with local government, acting as urban laboratories to develop tools and technologies that are then made available to all PTI members. UCEnvTF also fosters city- and county-based commercialization ventures. The *Environmental Monitoring Project*, for example, is a cooperative agreement between PTI and EPA to improve the use of environmental monitoring to enhance environmental management and public health. Another resource is environmental "SWAT" teams that help communities to quantify community economics for environmental cleanup, design a complete Environmental Management System (EMS), conduct ecological risk management and evaluate site characterization and remediation technologies for brownfield cleanup. It is also developing a climate-change tool kit for local governments.

SECTOR RELEVANCE environmental equipment, environmental services, local government

CONTACT

Ronda Mosley-Rovi
Director, Environmental Programs
Public Technology, Inc.
Tel: 202-626-2455
E-mail: mosley-rovi@pti.org

National Tribal Environmental Council

WEB SITE www.ntec.org

KEY WORDS air quality, hazardous waste, renewable energy, solid waste

PURPOSE to enhance each tribe's ability to protect, preserve and promote the wise management of air, land and water for the benefit of current and future generations.

SUMMARY Currently representing 108 member tribes, *NTEC* provides advocacy and technical services to all federally recognized tribes. NTEC promotes the development of environmental management capacity in a manner that respects each tribe's priorities and values. Current projects include:

FIGURE 9.4
Native Americans, who historically lived in harmony with nature and developed their own sustainable agriculture methods, are melding traditional teachings with present-day technologies to repair their lands following environmental degradation originating outside their borders (U.S. Army Environmental Center, with permission).

- The *Western Regional Air Partnership (WRAP),* which has an annual budget of $2 million and involves nearly 200 individuals from numerous tribes, ten western states, federal agencies and various stakeholder groups.
- The *Superfund/Hazardous Substances Project* is preparing a report on hazardous-waste issues that impact tribal governments.
- *Alaskan Resources* is an annotated bibliography of research on solid waste issues for villages and communities.

Various federal agency programs have special funding for environment-related projects on Native American lands. The Department of Energy, for example, has the *Renewable Energy Development on Tribal Lands,* a competitive grant program. To view the solicitation, see contact information.

NTEC publishes the *Tribal Vision* newsletter and maintains a *Resource Clearinghouse.*

MAIN RESOURCES OFFERED R1, R2, R4, R6, R12, R14

TECHNOLOGY FUNCTIONS T1, T2, T3, T10

SECTOR RELEVANCE tribal councils

CONTACTS

NTEC
Tel: 505 242-2175

Air quality cooperative
Bill Grantham
E-mail: bgrantham@ntec.org

Superfund
Lisa Gover
lgover@ntec.org

Alaskan Resources, Drinking Water
E-mail: jpardilla@ntec.org

Renewable Energy Development on Tribal Lands Solicitation
www.golden.doe.gov/businessopportunities.html

Selected State Programs

CALIFORNIA ENVIRONMENTAL PROTECTION AGENCY http://www.calepa.ca.gov

NEW JERSEY DEPARTMENT OF ENVIRONMENTAL PROTECTION http://www.state.nj.us/dep/

DEPARTMENT ENVIRONMENT PENNSYLVANIA http://www.dep.state.pa.us/l

KEY WORDS land use, river basin, technology transfer, technology verification

PURPOSE the states carry out their own environmental laws and implement U.S.EPA mandates. In addition, some have taken on as part of their mission to develop and verify technology and to provide technical assistance.

SUMMARY *The California Environmental Technology Certification Program* (CalCert) is an independent, recognized scientific and engineering evaluation of an environmental technology's performance. Technology manufacturers and developers define their performance claims and submit supporting data. CalCert reviews the information (and, where necessary, may require additional testing) to verify claims. If proven, the technologies, equipment and products receive certification verifying their performance claims. The program is voluntary and self-supporting; participating companies pay the cost of evaluating and certifying claims.

The *Office of Innovative Technology and Market Development* (New Jersey) makes information about innovative environmental technologies available and encourages their commercialization and use. Resources include technical manuals, technology seminars and collaboration with the Interstate Technology & Regulatory Cooperation Workgroup (see entry elsewhere in this chapter).

The *Growing Greener Program* (Pennsylvania) invests to preserve farmland and protect open space, eliminate the maintenance backlog in state parks, clean up abandoned mines and restore watersheds and provide new and upgraded water and sewer systems. Types of grants include: community grants for local recreation, park and conservation projects; rivers grants to conserve and enhance river resources; and land trust grants for acquisition and planning of open space and natural areas that face imminent loss. Lands must be open to public use and priority is given to habitat for threatened species. Heritage parks grants preserve and enhance natural, cultural, historic and recreation resources and stimulate heritage tourism. Grants may require 25–50% matching funds and are open to municipalities. Heritage grants are open to nonprofit land trusts and conservancies.

MAIN RESOURCES OFFERED R1, R3, R4, R5, R6, R12

TECHNOLOGY FUNCTIONS Various

SECTOR RELEVANCE environment equipment, municipal governments, nonprofits

CONTACTS

California Environmental Technology Certification Program
Frank S. Simpson/Roni Java
Tel: 916-341-6300

Public Affairs Office
http://www.calepa.ca.gov/CalCert/default.htm

Office of Innovative Technology & Market Development
PO Box 409
Trenton, NJ 08625-0409
Tel: 609-984-5418
Fax: 609-292-7340
E-mail: mwinka@dep.state.nj.us

Growing Greener Grants
Department of Conservation and Natural Resources
Apply to regional offices listed at:
http://www.dcnr.state.pa.us/growinggreener/gg-grants.htm

10

NGOS, Nonprofits and Professional Societies

Engineering Societies

American Society of Civil Engineers

WEB SITE www.asce.org

American Society for Engineering Education

WEB SITE www.asee.org

KEY WORDS design for the environment, energy efficiency, industrial technology, sustainable development, water quality

PURPOSE to apply sustainability principles in engineering practice and education

SUMMARY *ASCE* and *ASEE* engage in promoting sustainability via several forums. Among their initiatives on sustainability that seek to accelerate the transfer of innovation into practice accessible via links through their Web sites are:

- *Center for Sustainable Infrastructure,* established by the Civil Engineering Research Foundation (CERF), an affiliate of ASCE.
- *International Institute for Energy Conservation (IIEC),* an affiliate of CERF
- *Environmental and Water Resources Institute,* a semi-autonomous institute within ASCE, dedicated to serving its members and the public in the protection of natural resources and the enhancement of human well-being
- *Engineers Forum for Sustainable Development (EFSD),* sponsored by ASCE and ASEE, an interdisciplinary forum for exchange of information, identifying information on engineering education programs that incorporate sustainable development and encouraging engineering participation in sustainable development programs and activities at local, regional and national levels.

MAIN RESOURCES OFFERED R1, R2, R3, R12, R14

TECHNOLOGY FUNCTIONS Various

SECTOR RELEVANCE Various

CONTACTS

EWRI
E-mail: www.ewrinstitute.org

Autumn Richter, Director
arichter@asce.org

CERF
2131 K Street, NW
Suite 700
Washington, DC 20037
Tel: 202-785-6420
Fax: 202-833-2604

EFSD
www.asee.org/neic/efsd/default.cfm
Harry Tollerton
h.tollerton@aseee.org

Global Environmental Management Initiative (GEMI)

WEB SITE http://www.gemi.org

KEY WORDS climate change, design for the environment, greenhouse gases, materials substitution, water management

PURPOSE to foster environmental, health and safety excellence worldwide through the sharing of tools and information to allow business to help business achieve environmental excellence.

SUMMARY *GEMI* works with its member companies to develop and promote e-tools and other effective mechanisms to help businesses enhance their environment, health and safety (EHS) performance and extend and leverage the capabilities of EHS professionals in the companies. Its activities include benchmarking programs (e.g., Design for Environment and Water Sustainability: Water Use, Issues and Management), Web-based tools, an annual conference and partnerships with nongovernmental organizations (NGOs), government agencies and other industry groups. GEMI makes its tools available free to other businesses and the public via its Web site and publications. Its www.businessandclimate.org Web site has templates that help a company or facility develop an action plan for reducing greenhouse gas emissions. The publication *New Paths to Business Value: Strategic Sourcing–Environment, Health and Safety* addresses the business value of managing EHS in procurement. Another publication, *Environment: Value to the Top Line* helps companies identify new processes and ways of thinking about the role of environmental initiatives that can give them a competitive advantage (see Sample Project).

MAIN RESOURCES OFFERED R1, R4, R5, R10, R11, R14

TECHNOLOGY FUNCTIONS T1, T4, T5, T6, T7, T9

SECTOR RELEVANCE manufacturing

CONTACT

One Thomas Circle, NW
Tenth Floor
Washington, DC 20005
Tel: 202-296-7449
Fax: 202-296-7442
E-mail: info@gemi.org

Sample Project: Koch Petroleum Group

KEY WORDS air quality, sulfur dioxide.

ENVIRONMENTAL CHALLENGE Koch Petroleum Group was facing a challenge of complying with a new EPQ Clean Air Act standard for low sulfur gasoline by the year 2006.

PROJECT OBJECTIVES to create a cleaner fuel and meet the quality of its competitors at a lower price, while, at the same time, increasing the company's market share. Koch saw an opportunity to be the first company to introduce a low-sulfur gasoline 6 years before the mandated date.

PROJECT RESULTS To meet the competition in the market, where gasoline is purchased on the basis of price, the company partnered with one of its main customers, Holiday Stationstores, a chain of convenience stores across 11 northern states. Together, they launched Koch's Blue Planet™ with an advertising campaign to increase public awareness of the benefits of low-sulfur fuel. In the first year, Koch reported Holiday's market share of gasoline sales increased. About a fifth of Holiday's customers cited "product better for air" as a reason for their purchase.

SECTOR RELEVANCE petroleum

CONTACT See GEMI site, Member Companies for link to Koch.

Pew Center on Global Climate Change

WEB SITE www.pewclimate.org

KEY WORDS climate change, greenhouse gases

PURPOSE to educate the public and key policy makers about the causes and potential consequences of climate change and to encourage the domestic and international community to reduce emissions of greenhouse gases.

SUMMARY The *Pew Center on Global Climate Change,* a nonprofit, nonpartisan, independent organization, reports on environmental impacts, economics and policy concerning greenhouse emissions. It also fosters public dialogue through conferences and coordinates industry and government discussions on climate change. The center's *Business Environment Leadership Council* brings together all sectors of society to address climate change. The 37 participating companies serve in an advisory role, but provide no financial contributions. The reports online include: *Solutions Series,* providing practical tools to evaluate and reduce greenhouse gas emissions; *Economics Series* on large-scale macroeconomic models for assessing the costs associated with climate change policies; *Environmental Impacts Series*; and *Policy Series* on effective and equitable policy alternatives both in the

United States and abroad. Future topics will include greenhouse-gas emission and reduction verification, alternative reduction approaches and individual case studies.

MAIN RESOURCES OFFERED R4, R13, R14

TECHNOLOGY FUNCTIONS Various

Sector Relevance Various

CONTACT

Pew Center on Global Climate Change
2101 Wilson Blvd.
Suite 550
Arlington, VA 22201
Tel: 703-516-4146
Fax: 703-841-1422

Subscribe to E-mail list: http://www.pewclimate.org/forms/email_subscribe.cfm

Restore America's Estuaries (RAE)

WEB SITE http://www.estuaries.org

KEY WORDS coastal habitats, estuaries, wildlife refuge

PURPOSE to restore and protect America's estuaries and coastal lands before they disappear.

FIGURE 10.1
Estuaries vital to both wildlife and humans are targets of a nationwide restoration program.

SUMMARY *RAE* is an alliance of 11 regional coastal community-based environmental organizations. RAE, NOAA, National Ocean Service and National Marine Fisheries Service partner with state and federal agencies, nonprofit organizations, scientists and others. The RAE's *National Strategy to Restore Coastal Habitat: Restoring America's Estuaries*, aims to restore 1 million acres of estuary habitat by 2010. The Estuary Restoration Act, signed in 2000, authorizes $275 million over 5 years for restoration projects. Projects eligible for funding must have a non-federal sponsor. The maximum federal share of funding is 65%. *Funding for Habitat Restoration Projects: A Citizen's Guide* (found at www.restoration.nos.noaa.gov/htmls/resources/funding.html) provides a database on habitat restoration projects. Users can search by program, agency, funding support type and funding amount. The *Community-Based Restoration Program*, a partnership between RAE and NOAA, has committed $3.4 million for restoration of specific estuaries, including Long Island Sound, Chesapeake Bay and Louisiana's Mississippi Delta (see Sample Technology).

MAIN RESOURCES OFFERED R1, R2, R6

TECHNOLOGY FUNCTIONS T13

SECTOR RELEVANCE federal, state and local governments, nonprofits

CONTACT

Steve Emmett-Mattox
Program Director
Restore America's Estuaries
3801 N. Fairfax Drive, Suite 53
Arlington, VA 22203
Tel: 703-524-0248
E-mail: sem@estuaries.org

Sample Technology: Underwater Grass Restoration, Chesapeake Bay Foundation (CBF)

WEB SITE www.cbf.org

KEY WORDS coastal habitats, estuaries, wildlife refuge

ENVIRONMENTAL CHALLENGE Grass beds are vital natural buffers that filter pollution and calm wave energy to help sediment settle out. Their roots and rhizomes hold the substrate and the beds provide food for waterfowl and habitat for blue crabs, rockfish and speckled trout. In the Chesapeake Bay, on the coasts of Delaware, Maryland and Virginia, only about 12% of the original 600,000 acres of underwater grasses remain, however, because of excessive water pollution and sedimentation.

FIGURE 10.2
Dr. Robert Orth of the Virginia Institute of Marine Science examines a bed of eelgrass on a flat near an area where the Chesapeake Bay Foundation and VIMS conducted planting experiments, including with an experimental "paddlewheel" board that speeded up embedded seedlings in the river bottom

PROJECT OBJECTIVES to utilize a unique planting boat developed specifically for large-scale underwater grass restoration to test whether enough of the seedlings survive so that they naturally multiply.

PROJECT RESULTS CBF, partnering with the Virginia Institute of Marine Science (VIMS) and dozens of volunteers, planted 2 acres of eelgrass, making it the largest underwater grass restoration effort in the bay's history. The weeklong experiment restored an acre each of underwater grasses in two of Virginia's rivers that feed into the bay—the Rappahannock and the James. Instead of the painstaking traditional restoration method, underwater divers manually planting grass seedlings, the twin-pontoon boat is capable of replanting an acre per day. Grass seedlings are loaded onto nozzle-shaped tubes attached to two rotating metal wheels. When the wheels rotate the seedlings to the bottom of the river, air from a compressor injects them into the bottom sediment. CBF and the VMIS scientists are monitoring the results. VIMS scientists checking the grass beds nearly a year later have found evidence of restoration of grass beds. But they caution that this is likely due to a draught in the Maryland region that has contributed to a reduction in run-off that brings nitrogen pollution to the Bay

SECTOR RELEVANCE fisheries, recreation

CONTACT

CBF
Philip Merrill Environmental Center
6 Herndon Avenue
Annapolis, MD 21403
Tel: 410-268-8816

World Business Council for Sustainable Development (WBCSD)

WEB SITE www.wbcsd.ch/index.htm

KEY WORDS sustainable development

PURPOSE to provide business leadership for change toward sustainable development and to promote the role of eco-efficiency, innovation and corporate social responsibility.

SUMMARY *WBCSD* is a coalition of 150 international companies from more than 30 countries and 20 major industries. It incorporates a global network of more than 30 Business Councils for Sustainable Development (BCSDs)—national and regional councils and partner organizations in Africa, Asia, Europe, the Americas and Oceania. Case studies on the site are divided into seven categories: Innovation and Technology, Eco-Efficiency, Managing and Understanding Change, Dialogue and Partnership, Providing and Informing Customer Choice, Corporate Social Responsibility and Creating Sustainable Livelihoods (see Sample Projects after this entry). WBCSD takes part in four different types of projects. *Council Projects* develop new concepts and approaches to issues associated with business and sustainable development. *Member-Led Sectoral Projects* are conducted by companies within a given industry or sector that wish to work with each other on sustainable issues. Current projects are in the cement, electric utilities, forestry, mining and mobility industries and each has its own detailed Web site. *Advocacy and Awareness Projects* communicate WBCSD policy positions to opinion leaders. *Capacity-Building Projects* foster sustainable development practices across businesses, including education and training, stakeholder dialogues and information sharing. The quarterly journal *Sustain* contains case studies, policies and business practices. WBCSD partners with the International Chamber of Commerce on *Business Action for Sustainable Development (BASD)*, to facilitate partnering with other participants in sustainable development initiatives such as the World Summit on Sustainable Development.

MAIN RESOURCES OFFERED R1, R4, R5, R10, R12

TECHNOLOGY FUNCTIONS T6, T13

SECTOR RELEVANCE various, including cement, electric utilities, forestry

CONTACTS

WBCSD
4 Chemin de Conches
1231 Conches-Geneva
Switzerland
Tel: 41 22-839-3100

Fax:(41 22) 839 3131
E-mail: info@wbcsd.org

Business Action for Sustainable Development:
Eric Beynon
Tel: +33 1 49 53 30 65
E-mail: eric@basd-action.net
www.basd-action.net/

Back issues and subscription services for *Sustain* are available at http://www.sustain-online.org/.

Water Conservation and Reuse Program

Sample Project: General Motors de Mexico Web Site http://www.wbcsd.ch/casestud/index.htm

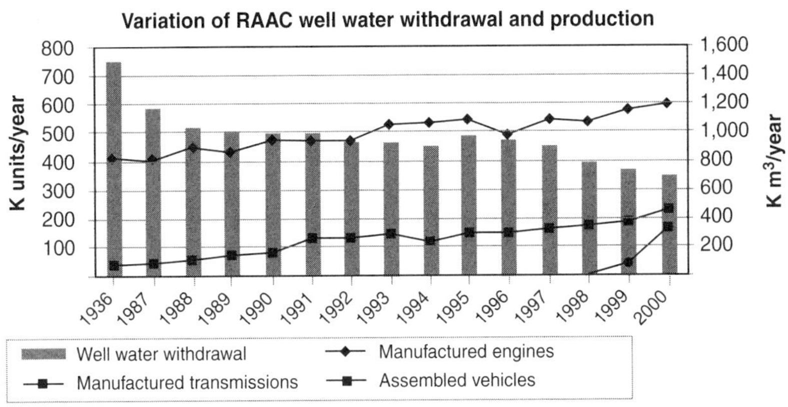

FIGURE 10.3
GMM was able to more than halve its well-water withdrawals while increasing production at its RHAC automobile manufacturing facility (courtesy WBCSD case study collection.

KEY WORDS aquifer, recycle/reuse, salinity, wastewater

ENVIRONMENTAL CHALLENGE With the opening of General Motors de Mexico (GMM) Ramos Arizpe Automotive Complex (RAAC) in 1980, production water requirements depended on a small semiconfined aquifer with a limited water storage capacity and a relatively high salt content (0.2%), which does not allow direct well-water use for industrial or domestic purposes. Since 1986, well-water levels have decreased while demand for high-quality water has increased with expansion of RAAC. By 2000, RAAC was manufacturing 590,000 engines, 171,000 transmissions and assembling 222,000 passenger vehicles.

PROJECT OBJECTIVES RAAC had three objectives: to secure water for production processes without depleting the aquifer, which was also used for local drinking water source; to desalinate the well-water supply; and to recycle and reuse wastewater.

PROJECT RESULTS GMM's Brine Recovery System (BRS) produces water with low salt content, used mainly by the engine and transmission plants, thus increasing the useable portion of water withdrawn from the well from 67 to 94%. By employing solar evaporation ponds, GMM avoids discharging salt-loaded water to a creek used for crop irrigation purposes. Total annual well-water withdrawal has dropped from 1,470,000 cubic meters in 1986 to 700,000 cubic meters in 2000. The average amount of well water needed to produce a vehicle fell from 32 cubic meters to 2.2 cubic meters.

SECTOR RELEVANCE automotive

CONTACT See main WBCSD page

Sample Project: Taiheiyo Cement Corporation, Using Urban Waste for "Eco-Cement"

WEB SITE www.wbcsd.ch/casestud/index.htm

KEY WORDS recycle/reuse, waste management

ENVIRONMENTAL CHALLENGE Japan generates more than 450 million metric tons of waste annually, but cannot construct new landfill sites due to land shortages and public resistance. While Japan uses up to 24 million metric tons of industrial waste a year for making ordinary cement, it has not utilized urban waste and incinerated ash.

PROJECT OBJECTIVES Taiheiyo Cement Corporation, one of Japan's leading national companies in the industry, undertook to develop technologies that utilized urban wastes such an incinerator ash and sewage sludge. The effort was designed to prolong the life of landfill sites in areas suffering waste-disposal problems and to remove toxic and harmful elements from its waste products.

PROJECT RESULTS Because of the high temperatures involved, a wide range of waste materials can be used in cement making without harming the environment. Taiheiyo's Eco-cement process destroys dioxins, and harmful heavy metals are recovered as useful nonferrous metal ores. The process also contributes to energy conservation and CO_2 reduction.

SECTOR RELEVANCE cement, waste management

CONTACT See Main WBCSD entry.

World Conservation Union (IUCN)

WEB SITE http://www.iucn.org

KEY WORDS biodiversity, forests, species, wetlands, wildlife

PURPOSE to influence, encourage and assist societies throughout the world to conserve the integrity and diversity of nature and to ensure that natural-resource use is equitable and ecologically sustainable.

SUMMARY *IUCN*, within the framework of global conventions, has assisted more than 75 countries to prepare and implement national conservation and biodiversity strategies.

IUCN brings together 78 states, 112 government agencies, 735 NGOs, 35 affiliates, and some 10,000 scientists and experts from 181 countries in a worldwide partnership.

A major project underway, the *Species Information Service (SIS,* a global species information source, is expected to be accessible via the Internet by early 2003*).* The SIS will expand upon and incorporate the IUCN "Red List" of threatened species (see below) and be continuously updated by its network of 7,000 experts. Another IUCN resource, the *Sustainable Use Specialist Group (SUSG)* operates through 17 regional networks (five in Africa, four in the Americas, three in Asia, and one each in Australia/New Zealand, the Pacific Islands, Europe, the Arctic, and the W. Indian Ocean Islands). Projects analyze and compare local use systems through case studies, regional reviews, workshops, and symposiums.

MAIN RESOURCES OFFERED R1, R2 R4, R10, R12, R15

TECHNOLOGY FUNCTIONS T1, T3

SECTOR RELEVANCE forestry, fisheries, government, NGOs, rangeland management

Sample Resource: Red List and Community Wetlands

The IUCN Species Survival Commission periodically issues its Red List of threatened species—"the world's most comprehensive inventory of the global conservation status of plant and animal species," For the first time, the latest (2000) Red List combines animals and plants into a single list containing assessments of more than 18,000 species (11,000 of which are threatened). Users include government agencies, wildlife departments, conservation-related NGOs, natural-resource planners, educational organizations, and others interested in slowing or halting the decline in biodiversity. Copies (price: £30, US$45) are available from IUCN Technical Publications (see Contacts).

Community Wetlands Management Plans —Uganda. A joint venture between the Uganda government and IUCN, the community planning approach to wetland management promotes multiple use management systems to enhance wetland benefits for local communities as well as offsite or downstream users. The Kanyabaha-Rushebeya community plan, for example, establishes a sanctuary offsite for the sitatunga, a rare wetland antelope, which is hunted by the local population because of the damage it causes to local farms. In addition, other features involve construction of footbridges to ease crossing of the wetland, promoting apiary activities, and boosting local fisheries.

CONTACTS

Species Information System

Dr. Mariano Gimenez Dixon

Tel: ++41.22.9990155

Fax: ++41.22.9990015

Email: mgd@hq.iucn.org

Uganda National Wetlands Programme

E-mail: mail@ugandawetlands.org

IUCN Technical Publications

IUCN Publications Service
Unit 219c Huntingdon Road
Cambridge CB3 0DL
United Kingdom
Tel: ++44 1223 277 894
Fax: ++44 1223 277 175
E-mail: info@books.iucn.org

Sustainable Use Team
IUCN USA Multilateral Office
1630 Connecticut Avenue, NW, Third Floor
Washington, DC 20009 USA
Tel: 202-518-2053
Fax: 202-387-4823
E-mail: sustainable@iucnus.org

World Resources Institute (WRI)

WEB SITE http://www.wri.org

KEY WORDS biodiversity, climate change, coastal ecosystems, forests, water quality

PURPOSE to provide, and help other institutions provide, objective information and practical proposals for policy and institutional change that will foster environmentally sound, socially equitable development.

SUMMARY *WRI* is an environmental think tank with three main goals: 1) to safeguard earth's climate from harm caused by greenhouse emissions, 2) reduce the use of materials and generation of waste in the production of goods and services and 3) guarantee access to environmental information and decisions regarding the environment. To supply this information, WRI manages *EarthTrends*, an environmental information portal divided into ten categories:

1. Coastal and marine ecosystems
2. Water resources and freshwater ecosystems
3. Climate and atmosphere
4. Population, health and human well-being
5. Economics, business and the environment
6. Energy and resources
7. Biodiversity and protected areas
8. Agriculture and food
9. Forests and grasslands
10. Environmental governance and institutions

Under each topic is a searchable database with data on 210 countries and 400 environmental variables spanning the last 100 years. WRI also has a large publications collection, including *World Resources 2001–People and Ecosystems: The Fraying Web of Life*. This booklet presents a comprehensive assessment of five of the world's main ecosystems: agro-ecosystems, coastal and marine, forest, freshwater and grassland. Other publications span topics ranging from maintaining biodiversity to international environmental law. While summaries of publications are available online, most must be purchased. WRI, working with British Petroleum, General Motors and Monsanto, published *Safe Climate, Sound Business: An Action Agenda*, a description of initiatives taken to reduce greenhouse-gas emissions while maintaining global economic prosperity. WRI has designed new energy-efficient office practices to reduce its greenhouse-gas emissions (for more information, WRI published a handbook entitled *WRI's Commitment to Reduce its CO_2 Net Emissions to Zero by 2005*).

MAIN RESOURCES OFFERED R1, R2, R14, R15

TECHNOLOGY FUNCTIONS T1, T3, T4, T13

SECTOR RELEVANCE Various

CONTACTS

World Resources Institute
10 G Street, NE
Suite 800
Washington, DC 20002
Tel: 202-729-7600
Fax: 202-729-7610

Publications:
Valerie Sheehan, Marketing & Database Manager
Tel: 202-729-7702
E-mail: valeriev@wri.org

To join E-mail list: http://www.wri.org/wri/email_list.html

Index

D

E

F

O

P

Q

R

S

U

V

W